Drinking Water in Sicily

Esterina Gagliano Candela ·
Giovannella Pecoraino · Sabina Morici ·
Leonardo La Pica
Editors

Drinking Water in Sicily

Hydrogeological and Hydrogeochemical Aspects

Springer

Editors
Esterina Gagliano Candela (Deceased)
Istituto Nazionale di Geofisica e
Vulcanologia, Sezione di Palermo
Palermo, Italy

Giovannella Pecoraino
Istituto Nazionale di Geofisica e
Vulcanologia, Sezione di Palermo
Palermo, Italy

Sabina Morici
Istituto Nazionale di Geofisica e
Vulcanologia, Sezione di Palermo
Palermo, Italy

Leonardo La Pica
Istituto Nazionale di Geofisica e
Vulcanologia, Sezione di Palermo
Palermo, Italy

ISBN 978-3-031-60472-0 ISBN 978-3-031-60473-7 (eBook)
https://doi.org/10.1007/978-3-031-60473-7

This Springer imprint is published by the registered company Springer Nature Switzerland AG
The registered company address is: Gewerbestrasse 11, 6330 Cham, Switzerland

This book is dedicated to Dr. Esterina Gagliano Candela (1953–2020), the creator and co-author of this book, with whom I have had the honour and privilege of sharing not only twenty years of work experience, but also a much longer deep and affectionate friendship. Without her commitment, patience and professionalism, this publication would not have seen the light of day. With the regret that she cannot be here with me to leaf through, with pleasure and with her characteristic critical sense, the result of our common work.

Giovannella Pecoraino

Preface

One of the most important issues in the environmental field is the protection of water resources and, in particular, the safeguarding of drinking water. Water resources need a careful prevention of pollution, rooted in multidisciplinary studies that increase the knowledge of the territory (geology, hydrogeology, geochemistry, geophysics) to highlight the vulnerable areas and to plan, where necessary, monitoring activities for the control and protection of the resource.

Our editorial project consists in a hydrogeochemical study of the sources in Sicily, southern Italy, used for drinking water purposes. The reference is the *Piano Regolatore degli Acquedotti* (PRGA) of Sicily (updated in November 2010), which lists all the springs, subdivided by province (a total of 1014), currently used to supply water to the Sicilian aqueducts. With regard to their geochemistry, the studies carried out by *Istituto Nazionale di Geofisica e Vulcanologia, Sezione di Palermo* (INGV-Pa hereafter) for the "Water Protection Plan for Sicily (2004–2005)" were used as our starting point.

The study of the chemical characteristics of groundwater is of fundamental importance from a hydrogeological point of view as it allows to deduce the origin of the water and the reconstruction of its evolution. Springs as a groundwater resource are paramount in relation to the availability of water resources due to the general decrease in precipitation.

In the first part of our project, all the existing data (flow rates, geochemical parameters, water quality, etc.) were collected, and a database was created on Excel sheets, which was then transferred to GIS software, after appropriate evaluations on the validity of the acquired data. Considering that the subject is quite delicate and involves various aspects (compliance with regulations, vulnerability of groundwater, anthropogenic pollution, etc.), carrying out a purely scientific study allows groundwater to be placed in a broader context than that of legislation aimed exclusively at compliance with the regulations in force.

Groundwater is part of the earth's water circulation system known as the hydrological cycle. Groundwater is derived from precipitation and surface water recharge; water infiltrates into the ground directly from precipitation, is recharged by streams and other natural water bodies, and is artificially recharged by man-made actions.

Palermo, Italy

Esterina Gagliano Candela
Giovannella Pecoraino
Sabina Morici
Leonardo La Pica

Acknowledgments There are many colleagues we wish to thank for collaborating on research project related to the book: Sergio Bellomo, Claudio Scaletta, and Manfredi Longo. In particular, we thank Dmitri Rouwet for reading the draft of the book and providing helpful comments. We thank also all authors who permitted us to use their figures. Finally, we thank the Istituto Nazionale di Geofisica e Vulcanologia—Dipartimento Ambiente for supporting the research published in this volume.

Contents

1 Regulatory Framework—Water for Drinking Purposes

Esterina Gagliano Candela, Giovannella Pecoraino, Sabina Morici, and Leonardo La Pica

Drinking water must meet the requirements for human consumption and includes all water distributed through public or private aqueducts, but also in bottles, tanks and other containers, used for domestic purposes, in the food industry and for the preparation of food and beverages. As such, the definition of drinking water does not depend on the origin of the water, but on the use of the resource, and therefore on the quality requirements it must guarantee in order to protect the health of its users.

The quality of water intended for drinking is regulated by a number of regulations, most of which have been transposed by EU directives. The regulations issued in Legislative Decree 31/2001 define "water intended for human consumption" as:

- treated or untreated water intended for drinking purposes, for the preparation of food and beverages, or for other domestic uses, regardless of its origin, supplied to users through a distribution network, tanks, bottles or various containers;
- water used by food businesses, for the manufacturing, treatment, storage or placing on the market of products or substances intended for human consumption, except for water which does not affect the wholesomeness of the final foodstuff.

E. Gagliano Candela (Deceased) · G. Pecoraino (✉) · S. Morici · L. La Pica
Istituto Nazionale di Geofisica e Vulcanologia, Sezione di Palermo, Via Ugo La Malfa 153, 90146 Palermo, Italy
e-mail: giovannella.pecoraino@ingv.it

S. Morici
e-mail: sabina.morici@ingv.it

L. La Pica
e-mail: leonardo.lapica@ingv.it

E. Gagliano Candela et al. (eds.), *Drinking Water in Sicily*,
https://doi.org/10.1007/978-3-031-60473-7_1

Table 1.1 shows the limit values (physical–chemical and inorganic-chemical) for the main global regulations governing water for (human) drinking use. The European directives are in accordance with the recommendations of the Food and Agriculture Organization (FAO) of the United Nations. The quality of drinking water is assessed based on measurements of concentrations of 64 parameters, which, for water to be potable, must be lower than the limit values known as "parametric values".

In Italy, the reference legislation for the protection of water from pollution is Legislative Decree no. 152 of 3 April 2006, PART THREE, SECTION II, on "Environmental regulations", published in the *Gazzetta Ufficiale* no. 88 of 14/04/2006 and entered into force on 29 April 2006. The regulation, which transposes Directive 2000/60/EC, describes the general procedures governing the monitoring and classification of the quality status of surface and groundwater.

Directive 2006/118/EC complements Framework Directive 2000/60/EC by adding criteria for the assessment of the good chemical status of water bodies, the identification and reversal of pollution trends, and the limitation of indirect discharges. Legislative Decree 30/2009 implements Directive 2006/118/EC, supplementing the provisions of "Part Three" of Legislative Decree 152/2006, and defines the "specific measures to prevent and control pollution and depletion of groundwater". Decree 260/2010 introduces the technical tools necessary for the effective implementation and enforcement of EU directives for the monitoring and classification of surface and groundwater bodies. This decree fully replaces Annex I to Part III of Legislative Decree 152/06.

1.1 Geology of Sicily

The Sicilian territory has varied morphological characteristics: the northern part of the island is predominantly mountainous with the reliefs (from west to east) of Monti di Trapani, Monti di Palermo, Monti di Trabia, Madonie, Monti Nebrodi and Monti Peloritani, while in the central-southern and south-western part of the island the landscape is characterised by modest reliefs with a typical hilly morphology with the exception of Monti Sicani. In the eastern part Sicily is dominated by the presence of the volcanic edifice of Etna, and in the south-eastern end by the Hyblean Highlands.

The structural geology of Sicily is characterised by three main tectonic elements (Fig. 1.1):

- ***the Avampaese (Foredeep) that occupies the south-eastern sector of the island*** and coincides with the Ibleo-Ragusano plateau, which is poorly deformed and towards which the various rocky bodies constituting the chain move. It is composed of a thick sequence of Meso-Cenozoic carbonate rocks and is affected by systems of direct faults that are sometimes attributed a transcurrent character.

Table 1.1 Limit values of inorganic chemical parameters for water intended for human consumption

Parameter	Unit	D.Lgs 31/01	Limit	UE 98/83/CE	WHO 2008	FAO Rev. 1997	U.S. EPA 2009
Electric conductivity (20 °C)	μS/cm	2500		2500			
pH		6.5–9.5		6.5–9.5			6.5–8.5
Hardness	°F	15–50					
Alluminium (Al)	μg/l	200	200	200	200		50–200
Ammonium (NH_4^+)	mg/l	0.5		0.5			
Antimonium (Sb)	μg/l	5	5	5	20	5	6
Silver (Ag)	μg/l		10				100
Arsenic (As)	μg/l	10	10	10	10	10	10
Barium (Ba)	μg/l				700	700	2000
Berillium (Be)	μg/l		4				4
Boron (B)	μg/l	1000	1000	1000	500		
Cadmium (Cd)	μg/l	5	5	5	3	3	5
Free cyanides (CN)	μg/l	50	50				
Clorine (Cl^-)	mg/l	250		250	250		250
Cobalt (Co)	μg/l		50				
Total Cromium (Cr)	μg/l	50	50	50	50	50	100
Cromium (VI)	μg/l	5	5				
Iron (Fe)	μg/l	200	200	200			300
Florine (F^-)	mg/l	1.5	1.5	1.5	1.5		2
Phosphorus (P)	mg/l			5			
Manganese (Mn)	μg/l	50	50	50	400	500	50
Mercury (Hg)	μg/l	1	1	1	1	1	2
Molibdenum (Mo)	μg/l				70		
Nichel (Ni)	μg/l	20	20	20	20	20	
Nitrate (NO_3^-)	mg/l			50	50	50	100

(continued)

Table 1.1 (continued)

Parameter	Unit	D.Lgs 31/01	Limit	UE 98/83/CE	WHO 2008	FAO Rev. 1997	U.S. EPA 2009
Nitrite (NO_2^-)	μg/l		500	500	200/3000	20	1000
Lead (Pb)	μg/l	10	10	10	10	10	15
Cupper (Cu)	μg/l	1000	1000	2000	2000	1000	1300
Dry Residue at 180 °C	mg/l	1500					
Selenium (Se)	μg/l	10	10	10	10	10	50
Sodium (Na)	mg/l	200		200			
Sulfate (SO_4^{2-})	mg/l	250	250	250			250
Tallium (Tl)	μg/l		2				2
Uranium (U)	μg/l				15		30
Vanadium (V)	μg/l	50					
Zinc (Zn)	μg/l		3000				5

- ***the Avanfossa (Foreland) represented by the Castelvetrano, Caltanissetta and Gela-Catania basins*** is located on the front of the mountain chain that constitutes a wide depression into which the products deriving from the dismantling of the progressively uplifting chain flow. This sector consists of carbonatic sequences (Triassic-Miocene), slightly deformed and stratigraphically covered by sediments of residual basins of a Plio-Pleistocene foreland;
- ***the mountainous chain***, with a southern verge, consisting of a series of tectonic units, resulting from the deformation of the original Meso-Cenozoic domains (Sicilide, Imerese and Sicano), superimposed one on the other, from west to east. The Sicilian Chain located in the northern sector, from Trapani to Messina, is represented by geological bodies with different characters and lithology, superimposed to form a complex stack of overlapping tectonic scales, up to more than 15 km thick, whose vergence develops towards the south and south-east. The various tectonic units have reached their currently visible reciprocal relationships over the last 20 million years as a result of a collision-related compression between Africa and Europe, which is probably still active today.

Three groups of layers can be recognised in this chain, which are the result of the collision of the North African plate with the South European plate. The first group of strata (geometrically higher) emerges in the Peloritanian sector and in the Calabrian Arc and is the result of the deformation of the southern continental margin of the Tethys (according to some authors, partly of the northern one as well). A distinction is made between Hercynian basement tectonic units (granites and metamorphic rocks of high and low grade) covered tectonically, in turn, during

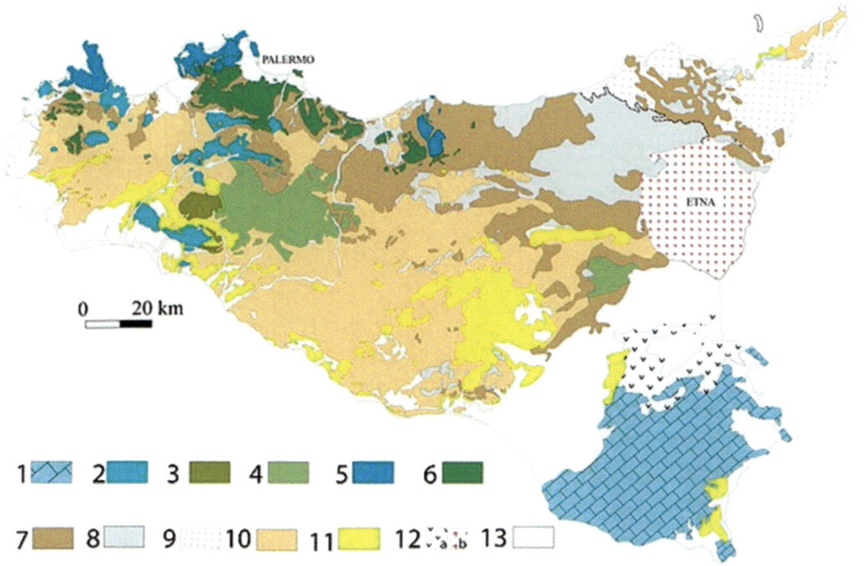

Fig. 1.1 Structural map of Sicily (modified from Catalano and D'Argenio 1982; Catalano et al. 1996). *Legend* (1) Iblean Avampaese Unit; (2) Carbonate-Pelagic Platform Unit (Trapanese-Saccense); (3) Monte Genuardo Unit (platform-basin transition; (4) Deep Sea Unit (Sicano); (5) Carbonate Platform Unit (Panormide); (6) Escarpment-Basin Unit (Imerese-Prepanormide); (7) Miocene flysch units (Numidian and internal flysch); (8) Sicilian units; (9) Calabro-Peloritanian crystalline units; (10) Mio-Pliocene synorogenic deposits; (11) Plio-Pleistocene syntectonic deposits; (12) Plio-Quaternary volcanites; (13) Pleistocene deposits

the high Palaeogene by Hercynian basement units with thin carbonate and clastic sedimentary cover (Trias-Oligocene).

The second group, occupying intermediate levels in the structural edifice, consists of rift strata, known as 'Variegated Clays' and derived from internal palaeogeographic domains (Sicilids). Their carbonate and terrigenous soils (Upper Jura–Lower Miocene) were placed between the Lower and Middle Miocene on the soils of the outermost units.

The third group is represented by tectonic units consisting of carbonate, silico-carbonate and silicoclastic sediments (Triassic-Miocene). This sequence also tore away from the original sedimentary substrate or with it, and tectonically stacked during the post-Miocene phases of compression.

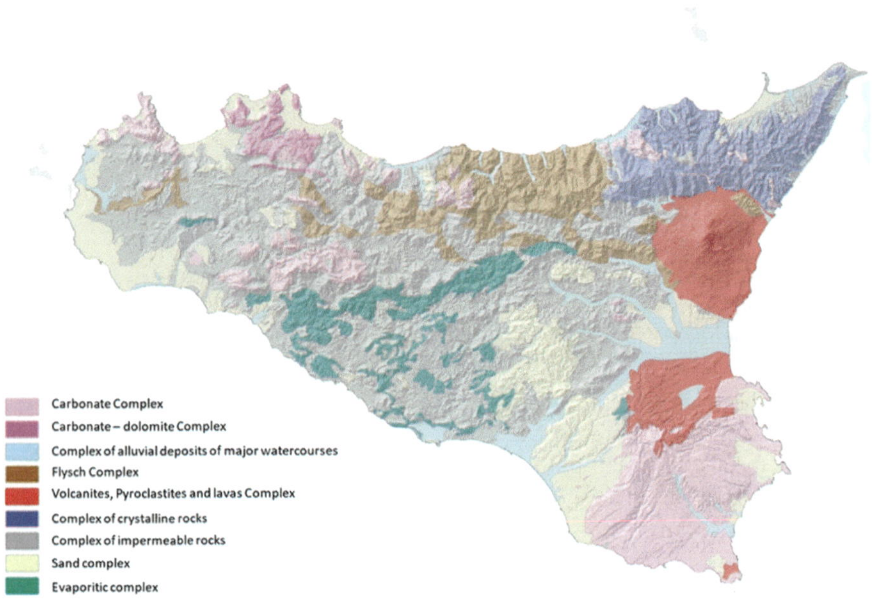

Fig. 1.2 Hydrogeological complexes

1.2 Hydrogeology of Sicily

1.2.1 Hydrogeological Complexes

Geological units can be grouped into hydrogeological complexes (Fig. 1.2) consisting of one or more lithological terms with similar hydrogeological characteristics (hydrogeological structure, permeability, porosity, infiltration capacity, vulnerability, hydrochemical facies).

Metamorphic complex (Paleozoic)
Composed mainly of phyllites and semi-schists, rocks that are poorly permeable to impermeable if they are not very fissured or altered, feeding a large number of springs of low flow. It outcrops in the north-eastern extremity of the island, in the Peloritani Mountains, where it is in contact with the flysch of the Nebrodi Mountains along the San Fratello-Taormina line.

Siliceous-calcareous argillitic complex (Middle Lias–Lower Miocene)
The lithotypes that compose this complex are represented by siliceous argillites with intercalations of 'resedimented biocalcarenites', calcilutites and marls, organogenic and glauconitic calcarenites. From a hydrogeological point of view, together, these rocks constitute an impermeable complex, forming the 'cover' of the underlying carbonate units of Trapani.

The rocks belonging to this complex outcrop in an E-W direction, in the northwestern part of the island between the Panormide carbonate structure of Mt Sparacio-Mt Monaco to the north, the Neogene evaporitic-carbonate terrigenous units to the south and the outcropping "horst" of the Trapani units to the east.

Carbonate Complex (Upper Triassic–Lower Miocene)
This hydrogeological complex is composed of calcareous rocks with a dolomitic and/or calcareous-dolomitic base and frequent intercalations of marly, calcareous-marly, siliceous and/or flinty, calcarenitic layers (Iblei Mts, Ragusano and some inland mountainous areas of Palermo and Trapani), or without interlayers (some units of the Syracusan area: Mt Climiti, Iblei di Noto, Capo Passero; the Mounts of Palermo, the massif of Mt Sparacio, the mountains of Corleone and Sciacca).

These rocks, which have a high relative permeability due to cracking and karstification, form an aquifer of considerable consistency.

Flysch complex of Permo-Triassic-Oligo-Miocene age
The different constituents of this complex are (1) largely allochthonous flyschoids (Sicilids, Varicolored Scaly Clays) composed of marls and clays prevailing on a stone fraction (limestones, limestone-marly limestones, sandstones ...); (2) marly-clayey and marly arenaceous flyschoid complexes, with evaporitic type rocks. Their outcrops are mostly located in the central and northern part of the island (M. Nebrodi, Madonie, Enna, the provinces of Caltanissetta and Agrigento, a large part of Trapani and the Valle del Belice); and (3) Predominantly clayey complexes of central-southern Sicily (hinterland of Gela, Ragusano, M. Erei etc.).

The latter complexes include the Flysch of M. Soro (Titonic-Cretaceous sup.) consisting of clayey-calcareous-arenaceous lithotypes and quartzarenites in large banks with clayey interlayers; silicid "Varicoloured Clays" with intercalations of marly limestone (Cretaceous-Eocene) passing upwards to clayey and marly-arenaceous oligocene alternations (Flysch of Troina-Tusa); clays and silty marls, quartz sandstones with clay interlayers and Oligo-Miocene quartzarenites (Numidian Flysch); greywackes, siltstones, clays with conglomerate and Oligo-Miocene arenaceous levels (Reitaino Flysch and Capo d'Orlando Flysch). The predominantly clayey and marly base terms are impermeable while the stone fraction is often intensely fractured and highly permeable. The predominantly arenaceous complexes are poorly to moderately permeable due to cracking and porosity.

Clayey-sandy and conglomeratic complex (middle to upper Miocene)
Present in the central-southern sector of Sicily, with considerable thickness (up to 1000 m) on the outcrop and in the subsoil (Lippone-Mazara area). From bottom to top, it is made up of marly clays, sandstones, sands and conglomerates (Un. Terravecchia Unit) on which cliff limestones are found (Messinian cliff complex, Baucina Unit) passing laterally to marls and marly clays. The hydrogeology of this domain is quite complex due to the existence of impermeable clay levels in between permeable, conglomeratic arenaceous and carbonate levels.

Complexes of evaporitic and marly-limestone deposits (Messinian-Inf. Pliocene)
This complex is mainly present in the central southern sector of the island; they form the Castelvetrano and Caltanissetta basins in complicated geometric relationships with the limestone units of the Sicanian and Saccenta domains. The lithotypes that compose them, belonging to different sedimentary cycles (late Miocene terrigenous and carbonate deposits and Messinian and Pliocene evaporitic and carbonate-marly deposits), form a sequence of hydrological complexes of low or very low permeability. The water table is located in the highly permeable levels of the vacuolar limestones and chalks of the Sicilian gypseous-sulphur series. The springs are very small and almost always contain mineralised water.

Sandy-calcarenitic complex (Middle Pliocene-Pleistocene)
This complex contains detrital-organic and terrigenous deposits (calcarenites, conglomerates, sands with clay intercalations). The permeability varies from good to worse depending on the degree of cementation of the calcarenite levels. It is mainly present in the western sector of the island and in particular along the coastal strips. The thickness varies from a few metres to about 100 m (Vittoria basin) and tends to become thinner in the south-western coastal areas. The coarser detrital levels may contain moderately thick aquifers that do not appear to interfere with the underlying deep Mesozoic carbonatic aquifers.

Volcanite Complex

Etna basic volcanite complex (Upper Miocene-Holocene)
The volcanic products that make up the volcanic edifice of Etna and the northern margin of the Iblei (Lentinese) are generally lavas and pyroclastites characterised by high permeability due to cracking and porosity, while the self-cemented volcanoclastic products (lahars) have low permeability.

Lavas and volcanoclasts, resting on a poorly permeable sedimentary base, form a large "suspended" and cold aquifer, generally not in communication with the deeper aquifers. Interaction with the deep circulation can occur through all those volcanic or volcano-tectonic structures (chimneys, craters, domes, large fracture systems, etc.) that connect the surface with the deep structures.

Acidic volcanites of the minor islands (Pleistocene-Oligocene)
Also in this case, the potentially water-bearing units are volcanic products generally characterised by good permeability (lavas, ignimbritic deposits, etc.) interspersed with more or less thick horizons of cemented and clayey volcanoclastic materials with low permeability. The permeable horizons can form aquifers, sometimes characterised by high temperatures (Aeolian Islands, Pantelleria) and the volcanoclasts form more or less continuous impermeable septa that determine the superimposition of aquifers with different chemical and physical characteristics.

Table 1.2 Hydrogeological Basins list

Hydrogeological basins
Madonie Mountains
Mount Etna
Iblei Mountain
Palermo Mountains
Sicani Mountains
Nebrodi Mountains
Peloritani Mountains
Trabia-Termini Imerese Mountains
Trapani Mountains
Rocca Busambra
Plain of Castelvetrano- Campobello di Mazara
Plain of Marsala—Mazara del Vallo
Piazza Armerina
Plain of Catania
Plain of Palermo
Caltanissetta Basin
Plain and Mountains of Bagheria
Plain of Gela
Plain of Licata

Recent deposits (Pleistocene)
These complexes are mainly represented by recent alluvial deposits, with maximum capacities in the order of a hundred cubic metres. Given their extreme granulometric variability, they represent multi-water table aquifers, separated by semi-permeable or impermeable levels.

1.3 Hydrogeological Basins

The territory of the Region has been subdivided into 19 hydro-geological basins (Table 1.2 and Fig. 1.3), which are in turn subdivided into 82 water bodies (Table 1.4 and Fig. 1.4) (INGV 2005) (Table 1.3).

1.4 The Sicilian Aquifers

In the Sicilian territory there are discrete aquifers in the eastern zone, in the volcanites of Etna and in the carbonate complexes of the Hyblean Montain. Aquifers of good local importance are present in the carbonate chain structures of the northern part of the island, from the Trapanese Mountains to the Peloritani Mountains.

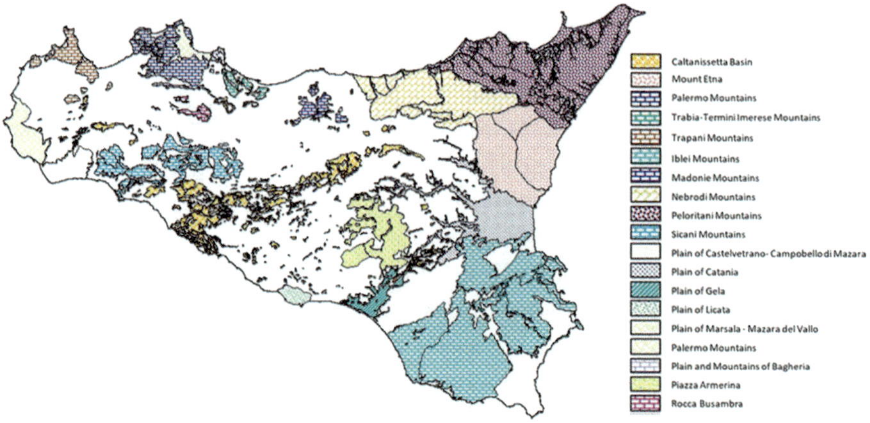

Fig. 1.3 Hydrogeological basins

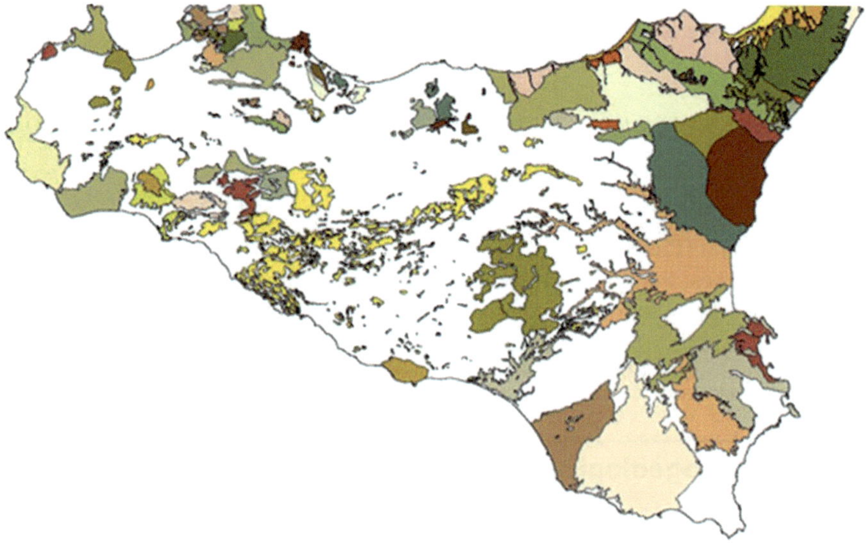

Fig. 1.4 Water bodies

The central part of the island is made up of marly-clayey-arenaceous complexes, which are impermeable and contain limited or scarcely productive aquifers of little hydrogeological interest.

Table 1.4 shows a classification of the Sicilian geological units on the basis of their permeability characteristics.

The Sicilian aquifers are listed and described below in order of their resource importance:

Table 1.3 Hydrogeological basins and water bodies considered in this work (modified from Pecoraino et al. 2008)

Hydrogeologic basin	Aquifer	Facies	Type of aquifer	Intrinsic vulnerability/land use
Etna Mount	Etna East	Quaternary volcanites	Unconfined	High/mixed
	Etna North			
	Etna West			
Palermo Mountains	Belmonte-Pizzo Mirabella	Mesozoic carbonate, fractured and karstfied, unconfined and confined by Oligo-miocene clays (Numidyan Flysch)	Unconfined and confined	High/mixed
	Castellaccio Mount			
	Cuccio Mount-Gibilmesi Mount			
	Kumeta Mount			
	Mirto Mount			
	Pizzo Vuturo-Pellegrino Mount			
Trabia-Termini Imerese Mountains	Capo Grosso-Torre Colonna	Mesozoic carbonate, fractured and karstfied, unconfined and confined by Oligo-miocene clays (Numidyan Flysch)	Unconfined and confined	High/mixed
	Pizzo di Cane-San Calogero Mount			
Madonie Mountains	Pizzo Carbonara-Pizzo Dipilo	Mesozoic carbonate, fractured and karstfied, unconfined and confined by Oligo-miocene clays (Numidyan Flysch)	Unconfined and confined	High/natural park

(continued)

Table 1.3 (continued)

Hydrogeologic basin	Aquifer	Facies	Type of aquifer	Intrinsic vulnerability/land use
Trapani Mountains	Bonifacio Mount	Mesozoic carbonate, fractured and karstfied, unconfined and confined by Oligo-miocene clays (Numidyan Flysch) and Tortonian Clays	Unconfined and confined	Very high/mixed
	Erice Mount			
	Ramalloro Mount-Inici Mount			
	Sparagio Mount-Monaco Mount			
Rocca Busambra	Rocca Busambra	Mesozoic carbonate, fractured and karstfied, unconfined and confined by Oligo-miocene clays (Numidyan Flysch) and Tortonian Clays		
Hyblean Mountains	Lentinese	Miocene-Pleistocene volcanites, Mesozoic carbonate, fractured and karstfied, unconfined and confined by Tortonian and Pliocene clays	Unconfined and confined	High/mixed
	Siracusano northwestern			
	Ragusano			
	Vittoria Plain	Plio-Pleistocenic alluvial coastal	Multi-layered	High/intensive agriculture, petrolchemical industry
	Augusta-Priolo Plain			

(continued)

Table 1.3 (continued)

Hydrogeologic basin	Aquifer	Facies	Type of aquifer	Intrinsic vulnerability/land use
Nebrodi Mountains	Pizzo Michele- Castelli Mount	Arenaceus fractured and porous unconfined	Unconfined	Middle/Natural Park, mixed
	Reitano-Castellaci Mount			
	Caronia	Alluvial coastal unconfined aquifer	Unconfined	High/intensive agriculture
	Santo Stefano			
	Tusa			
Peloritani Mountains	Alcantara	Arenaceous, metamorphic and carbonate fractured and porous unconfinate and confinate by Oligo-Miocenic clays (Numidyan Flysch) and Tortonian Clays	Unconfined and confined	Middle/Natural Park, mixed
	Eastern Peloritani			
	Southern Peloritani			
	South-eastern Peloritani			
	Mirto Mount			
	Naso			
	North-eastern Peloritani			
	North-western Peloritani			
	Western Peloritani			

(continued)

Table 1.3 (continued)

Hydrogeologic basin		Aquifer	Facies	Type of aquifer	Intrinsic vulnerability/land use
		Brolo	Alluvial coastal unconfined aquifer	Unconfined	High/intensive agricolture, petrolchemical industry
		Messina-Capo Peloro			
		Gioiosa Marea			
		Timeto			
		Roccalumenra			
		S. Agata—Capo D'Orlando			
		Barcellona—Milazzo Plain			
Sicani Mountains		Magaggiaro Mounts	Miocene-Pleistocene volcanites, Mesozoic carbonate, fractured and karstfied, unconfined and confined by Tortonian and Pliocene clays	Unconfined and confined	High/intensive agricolture: Vines, olive grove, citrus, firstling
		Montevago			
		Southern Saccense			
		Central Sicani			
		Southern Sicani			
		Easthern Sicani			
		Northern Sicani			
Catania Plain		Catania Plain	Alluvial coastal multi-layered	Multi-layered	High/intensive agricolture, industrial

(continued)

Table 1.3 (continued)

Hydrogeologic basin	Aquifer	Facies	Type of aquifer	Intrinsic vulnerability/land use
Castelvetrano-Campobello di Mazara Plain	Castelvetrano-Campobello di Mazara Plain	Porous coastal multi-layered	Multi-layered	High/intensive agricolture
Marsala-Mazara del Vallo Plain	Marsala-Mazara del Vallo Plain	Porous coastal multi-layered	Multi-layered	High/intensive agricolture
Piazza Armerina	Piazza Armerina	Porous, shallow unconfined	Unconfined	High/intensive agricolture
Palermo Plain		Calcarenites	Unconfined	High/villages
Plain and montains of Bagheria		Calcarenites and limestone	Unconfined	High/villages
Caltanissetta Basin	Caltanissetta Basin	Clays		
		Evaporitic limestone and gypsum Pliocenic and tortonian clays	Unconfined and confined	High/mixed
Licata Plain	Licata Plain	Alluvial coastal	Unconfined or locally confined	High/intensive agricolture
Gela Plain	Gela Plain	Porous coastal multi-layered	Multi-layered	High/intensive agricolture

Table 1.4 Classification and characteristics of Sicilian geological units (modified by Cassa per il Mezzogiorno 1976)

Classification of geological unit						
Age	Facies	Permeability				
		Very little or none	Low	Mediocre	Discrete	High
Recent	Alluvium, dune sand, clays	–	–	–	–	
Pleistocene and recent	Volcanites	–	–	–	–	–
Pleistocene and permeabile Pliocene	Limestone, sands, sandstone	–	–	–	–	
	Sands, gravel, clays	–	–	–		
	Limestones, sands	–	–	–	–	
Miocene—Pliocene	Chalky-sulphur and trub series	–	–	–		
Pliocene	Flysch and predominantly clayed sediments	–				
	Clays and marls					
	Clays					
Eocene-middle Miocene	Numidian Flysch					
Eocene-Helvetian	Flaky clays, marly limestones and sandstones of Peloritani M					
Upper Oligocene-Lower Miocene	Chaotic clays with sandstone and breccia intercalations					
Eocene	Marne					
	Carnic Flysch					
	Permian Flysch					
	Flysch and predominantly arenaceus sediments	–	–			

(continued)

Table 1.4 (continued)

Classification of geological unit						
Age	Facies	Permeability				
		Very little or none	Low	Mediocre	Discrete	High
Tortonian	Conglomerates and sands					
Oligo-Middle Miocene	Reitano sandstones					
Eocene-middle Miocene	Numidian Flysch					
Oligo-Lower Miocene	Sandstones of Capo D'Orlando					
Upper Titonic-Lower Cretaceus	Mt Soro sandstones					
Miocene	Palazzolo limestones	–	–	–	–	–
	Tellaro marls	–	–			
Mio-Pliocene	Volcanites associated with calcarenites	–	–	–	–	
Oligo-Miocene	Ragusano limestone complex	–	–	–	–	–
Mesozoic	Limestones, marls, dolomites and volcanites	–	–	–	–	
Paleozoic	Crystalline schists	–				

(1) ***Volcanites of Etna***

These volcanites have a very high permeability and constitute the highest relief in Sicily, where there is abundant rainfall (900–1200 mm), mostly snow. They represent the richest aquifer on the island.

(2) ***Miocene limestones of the Hyblean Mountains***

Composed of limestone from the Syracuse and Ragusa areas. They have a very high permeability and are located in areas of moderate rainfall (800–1100 mm). They contain an almost continuous aquifer (in the porous and/or fractured layers) drained by a karstic network that developed along fault lines. The water table flows into numerous important springs located in the deep valley incisions (quarries) that cut into the limestone.

(3) ***Mesozoic limestones***

For this unit the permeability is high, although lower than that of Miocene limestones because it is limited to fault and fracture zones and karst tunnels. These massifs, which are generally high in altitude, receive a fair amount of precipitation (900–1300 mm). The water table originates from numerous springs (particularly in the Monti di Palermo and Monti Sicani), or feeds the Plio-Pleistocene sands and calcarenites immediately in contact, or, to a lesser extent, is lost to the sea (Trapani and Cefalù areas). The aquiferous zone of the limestones is generally limited to the first 200–300 m of the unit (beyond this depth the permeability is zero due to the compactness of the rock).

To this group belong:

- the Mounts of Trapani
- the Mounts of Palermo
- the Mounts of Termini Imerese-Trabia
- Madonie Mountains
- Nebrodi Mountains
- the Sicani Mountains.

(4) ***Lentinese volcanites***

These aquifers develop in the Lentini Basin, also affect the adjacent basins, and have an average permeability. The feeding area lies between 200 and 600 m a.s.l. and precipitation is on average low (550–800 mm). The average thickness of the aquifer is about 200 m; towards the north, however, the thickness of the volcanites can sometimes exceed 500 m. Permeability is related to some facies of the volcanites (lapilli, breccias, fissured lavas, intercalated calcareous layers). The water table is partially drained by the hydrographic network; the springs have partly disappeared and the aquifer is now exploited by means of numerous drillings located in the Francofonte-Palagonia-Lentini area.

(5) ***Plio-Quaternary sands and calcarenites***

These aquifers essentially develop in the following areas:

- Vittoria-Caltagirone area
- Piazza Armerina-Mazzarino area
- coastal plains from Trapani to Sciacca
- coastal area from Castellammare del Golfo to Termini Imerese.

The small outcrops of Licata-Gela, Agira and Ragalbuto, Augusta, Siracusa and Avola should also be mentioned.

From a morphological point of view, the calcarenite areas are relatively low-lying plains or plateaus (less than 160 m a.s.l.) subdivided into plates by the hydrographic network. The unit is fairly permeable. Its thickness varies from 50 to 100 m in the Vittoria basin, from 10 to 50 m in the Palermo area, and it does not exceed 100 m in the coastal plains from Trapani to Sciacca, while it can reach 200–250 m at Piazza Armerina and Leonforte. Precipitation in these areas is on average low (500–700 mm) but some calcarenitic plates also receive lateral feeding from the limestone aquifers (areas of Syracuse, Vittoria, Trapani, Palermo, Termini Imerese). The stratum is generally shallow and continuous, which explains the numerous catchment works that affect it, particularly in the Vittoria, Palermo, Termini Imerese and Mazara del Vallo areas. There is also some drainage of the water table by the rivers and an outflow towards the sea in the coastal areas. However, the amount of water still available in the aquifer appears to be small.

(6) ***Floods***

The various alluvial aquifers follow the course of the main river valleys. Due to their silty nature and the consequent reduced permeability, alluvial waters are of limited interest in the Trapanese area, along the Belice River and the Gela River. In addition, the salinity of the water considerably limits the possibility of using the alluvial aquifer of the Platani and Southern Imera.

On the other hand, the following aquifers are of reasonable interest:

- the alluvium on the northern side of the island
- the alluvium of the Simeto and the Plain of Catania
- the alluvium of the Messina torrents
- the ancient alluvium of the Acate
- the ancient alluvium of the Milazzo Plain.

The recent alluvium varies in thickness from a few metres to a maximum of 80 m (Piana di Catania). They include soils of varying permeability. The alluvium of the Fiumare of Messina is generally coarse and very permeable. In other areas there is often a layer of superficial clayey silts overlying a layer of sand and gravel. The ancient alluvium of Milazzo is made up of an alternation of gravels and sands with clayey-silt intercalations and a layer of sand. Milazzo's ancient alluvium is made up of alternating gravel and sand with clayey-silt interbeds over 200 m deep. These alluvial reservoirs are fed by the rivers with which the water table is in equilibrium.

(7) ***The Gypsum-Sulphur Series***

This series is present in discontinuous outcrops throughout central and western Sicily; it is composed of small sedimentation basins, hollowed out in a vast area of Miocene marls and clays. The series includes soils of very different nature (Tripoli, basic limestones, chalks).

The only horizons of hydrogeological interest are chalks and basic limestones. These two terms are encountered more frequently in the sector of Licata, Campobello di Licata, Racalmuto.

(8) ***Flysch sandstones***

Various types of sandstones are distinguished, depending on their origin and age, but their hydrological behaviour is practically identical. Sandstones are mainly developed in northern Sicily and, to a lesser extent, in western Sicily. Their permeability, generally low, is related to the disintegration of the upper part of the rock under the effect of atmospheric agents and its fracturing in relation to tectonic events. A series of small, independent permeable flaps are thus formed. These tiny reservoirs give rise to a multitude of tiny, often perennial springs.

An exception to this pattern is the Capo d'Orlando Flysch, whose aquifer is made up of basic conglomerates, about 50 m thick, extending over a large surface area with high rainfall. Important springs are present.

(9) ***The altered zone of the crystalline schists***

The schists appear exclusively in the Peloritani; if compact and unaltered they are not permeable. Nevertheless, secondary permeability exists in the altered zone of the rock, which can be up to 30–50 m thick. This permeability, however, is always limited and the amount of underground flow is small. The water collected and drained from the alteration aquifers flows into numerous springs with a generally low flow rate, but which together contribute to the supply of the torrents.

Only a part of these springs is used for aqueducts.

(10) ***Flysch and predominantly clayey sediments***

These are marly and clayey units of almost zero permeability which cover vast areas of northern, western and southern Sicily. They have no water resources except for very small water tables suspended in some detrital or arenaceous layers, or in small alluvial plates along the hydrographic network.

1.5 Hydrogeochemical Aspects of Sicily

The Sicilian territory has a geological diversity that has a decisive influence on the geochemistry of the groundwater. The chemical composition of water can be linked both to natural processes such as the leaching and dissolution of the minerals that make up the rocks with which the water interacts as it travels underground, and to anthropogenic processes such as pollution from industrial or agricultural activities. The majority of the strategic aquifers for drinking water use are karstic and therefore based on carbonate rocks. The Etnean volcanic edifice hosts another strategic aquifer, of significant potential with respect to the island's needs. There are also a series of small and medium-sized aquifers that are certainly not of strategic importance, but which, in many cases, manage to solve the supply problems of small centres. These aquifers are mainly located in arenaceous reservoir rocks or on minor carbonate outcrops. In the central-southern part of the island, the underground waters are practically unusable because they generally circulate in units belonging to the gypsum-sulphur series and are therefore of poor quality.

The Langelier-Ludwig classification diagram was used for the geochemical characterisation of the waters. This diagram, based on the solution content of the major constituents (Na, K, Ca, Mg, Cl, SO_4, HCO_3, CO_3), distinguishes waters into four sectors, according to anion and cation pairs (Fig. 1.6). Depending on their position within the diagram, waters are classified as:

- sulphate-chlorinated-alkaline (quadrant I);
- bicarbonate-alkaline (quadrant II);
- bicarbonate-earthy-alkaline (quadrant III);
- sulphate-chlorinated-earthy alkaline (quadrant IV).

Carbonate aquifers are strategically the most important from the point of view of drinking water supply. The hydrogeological characteristics of the carbonate rocks and limestone in particular, allow the infiltration in depth of most of the meteoric precipitations that occur in these areas, thus allowing the supply of aquifers of great potential that have the capacity to provide water of excellent quality from the point of view of drinking water use.

Their chemical connotation is of the bicarbonate-calcium type, due to the chemical composition of the limestone $CaCO_3$ and the dolomite $CaMg(HCO_3)$, and they fall in quadrant III of the Langelier-Ludwig diagram, where the waters with Ca and Mg as the prevalent cationic pair and carbonates and bicarbonate as the prevalent anionic pair are located. The other major constituents are also present, but in lower concentrations.

The electrical conductivity of these waters is generally between 300 and 800 μS/cm. The prevailing location of occurrence of these aquifers is in the western sector of the island in correspondence with the "Sicilian Apennine-Maghrebide chain", in the Hyblean Plateau in the south-eastern part of Sicily, and in the Sicanian Mountains. Minor carbonate outcrops are found on the Nebrodi Mountains ridge and in the Peloritani Mountains (Fig. 1.5).

Another large strategic aquifer capable of producing significant quantities of water is located in the Etna region. On the one hand, the chemical characterisation of these waters is conditioned by the presence, throughout the area, of widespread carbon dioxide outgassing that dissolves in the aquifers in the form of bicarbonate, which is, in fact, the prevalent anionic species in the majority of Etna's aquifers. The cations, on the other hand, are somewhat more balanced with a slight prevalence of Mg and Ca. In the Langelier-Ludwig diagram, Etna's waters fall mainly within quadrant III of the bicarbonate-alkaline waters.

In significant portions of the Sicilian surface, flyschoid-type sediments outcrop, consisting mainly of alternating successions of clays and quartzarenites. The latter, which certainly do not have such a high permeability compared to that of carbonate rocks, are however the site of locally important aquifers. The compositional variability of these quartzarenites, together with the depth of circulation of the aquifer, influence the geochemical characterisation of these waters, which in

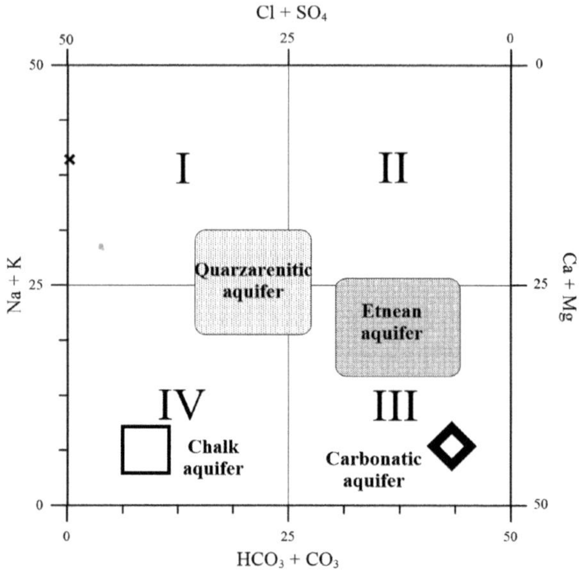

Fig. 1.5 Langelier Ludwig diagram. Modify by Favara and Gagliano Candela (2005)

the Langelier-Ludwig diagram fall within a wider range than the other aquifers. Their saline load is also extremely variable, ranging from tens of μS/cm (the Mt Catarineci aquifers near Geraci Siculo) to thousands of μS/cm (the aquifers in the Finale di Pollina coastal area). The chemical composition of these waters is characterised by the prevalence of the Cl-SO_4 pair over the HCO_3–CO_3 pair, and of Na and K over Ca and Mg. These waters, in the Langelier-Ludwig diagram (Fig. 1.6) fall in the centre of the diagram in an area that includes portions of the four quadrants.

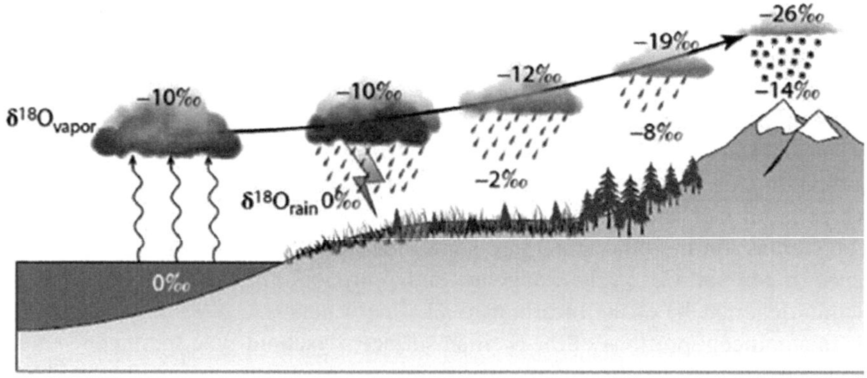

Fig. 1.6 $\delta^{18}O$ fractionation starting with the evaporation of seawater, condensation in clouds and subsequent repeated precipitation inland

In areas with evaporitic sediments, aquifers are strongly influenced by the high solubility of the reservoir rocks. In the gypsum aquifers, in the Langelier-Ludwig quadrant IV, selenitic waters with a calcium sulphate component circulate. These waters are of poor quality because, in addition to having a high Ca and SO_4 content, they have a high salinity which in some cases even compromises irrigation or industrial use.

Another typical characterisation of evaporites is the chlorinated-sodic one connected to water circulation in rock salt units. Table 1.5 shows the geochemical classification of waters in hydrogeological basins according to the Langelier-Ludwig classification.

1.6 Isotopic Composition of δ^2H and δ^{18}O of Waters (Groundwater + Rainwater) in Sicily

The isotopic composition of water represents the most useful tool for hydrogeological investigations. In a hydrological basin, rainfall is the main groundwater recharge. Consequently, groundwater has an isotopic connotation that reflects that of precipitation. In general, the δ^{18}O and δ^2H values of groundwater fall close to the local precipitation line and the average isotopic composition of groundwater is similar to the annual average of the isotopic composition of precipitation occurring in the catchment area, weighted by the monthly amount. Differences in isotopic composition of rainwater are caused by fractionation phenomena linked to altitude, latitude, season, continent and quantity effects (Fig. 1.6). More details on the isotopic composition of rain- and groundwater can be read in the IAEA (1981) report.

By exploiting the variations induced by the effect of altitude on the isotopic composition of rainfall, it is possible to estimate the vertical isotopic gradient of rainfall in a basin, making it possible to derive, as a function of the isotopic ratios of groundwater, the average elevations of infiltration (feeding) water within a basin. The geographic distribution of δ^{18}O in Sicily is shown in Fig. 1.7.

During the last decades, several articles on the isotopic composition of rainwater and groundwater in central Mediterranean area have been published (e.g. Longinelli and Selvo 2003; D'Alessandro et al. 2004; Grassa et al. 2006; Liotta et al. 2006a, b; Liotta et al. 2013 and references therein), but many concentrated on the modelling of some specific areas of the Sicilian territory.

Liotta et al. (2013) modelled for the first time the entire Sicilian area based on data of two hydrological years, determining the relation between rain- and groundwater. These authors computed the overall Weighted Local Meteoric Water Line (WLMWL) for Sicily ranging from $-10.5‰ < \delta^{18}O < -5.5‰$ and $-62‰ < \delta^2H < -29‰$, a slope of 6.75 and an intercept of 8.2. With the aim to construct a detailed, high resolution model of δ^{18}O and δ^2H relationships, Sicily was divided in five sectors in function of morphology and hydrogeological characteristics (Fig. 1.8 and Table 1.6), and local parameters were determined.

Table 1.5 Langelier-Ludwig water classification of Sicilian aquifers

Hydrogeological basin	Langelier-Ludwig water classification
Madonie Mountains	Bicarbonate-earthy alkaline
Mount Etna	Bicarbonate-earthy alkaline
	Bicarbonate-alkaline
	Chlorinate-suphate—alkaline
Hyblean Mountains	Bicarbonate-earthy alkaline
	Chlorinate-suphate—earth alkaline
	Chlorinate-suphate—alkaline
Palermo Mountains	Bicarbonate-earthy alkaline
	Chlorinate-suphate—earth alkaline
Sicani Mountains	Bicarbonate-earthy alkaline
	Chlorinate-suphate—earth alkaline
Nebrodi Mountains	Bicarbonate-earthy alkaline
	Chlorinate-suphate—earth alkaline
Peloritani Mountains	Bicarbonate-earthy alkaline
	Chlorinate-suphate—earth alkaline
Trabia-Termini Imerese Mountains	Bicarbonate-earthy alkaline
	Chlorinate-suphate—earth alkaline
Trapani Mountains	Bicarbonate-earthy alkaline
	Chlorinate-suphate—earth alkaline
	Chlorinate-suphate—alkaline
Rocca Busambra	Bicarbonate-earthy alkaline
Castelvetrano-Campobello di Mazara Plain	Bicarbonate-earthy alkaline
	Chlorinate-suphate—earth alkaline
Marsala-Mazara del Vallo Plain	Chlorinate-suphate—earth alkaline
Piazza Armerina	Bicarbonate-earthy alkaline
	Chlorinate-suphate—earth alkaline
Catania Plain	Bicarbonate-earthy alkaline
	Chlorinate-suphate—earth alkaline
	Chlorinate-suphate—alkaline
Palermo Plain	Bicarbonate-earthy alkaline
	Chlorinate-suphate—earth alkaline
	Chlorinate-suphate—alkaline
Caltanissetta Basin	Chlorinate-suphate—earth alkaline
	Chlorinate-suphate—alkaline
Plain and Mountains of Bagheria	Chlorinate-suphate—earth alkaline

(continued)

Table 1.5 (continued)

Hydrogeological basin	Langelier-Ludwig water classification
	Chlorinate-suphate—alkaline
Gela Plain	Chlorinate-suphate—earth alkaline
	Chlorinate-suphate—alkaline
Licata Plain	Chlorinate-suphate—earth alkaline
	Chlorinate-suphate—alkaline

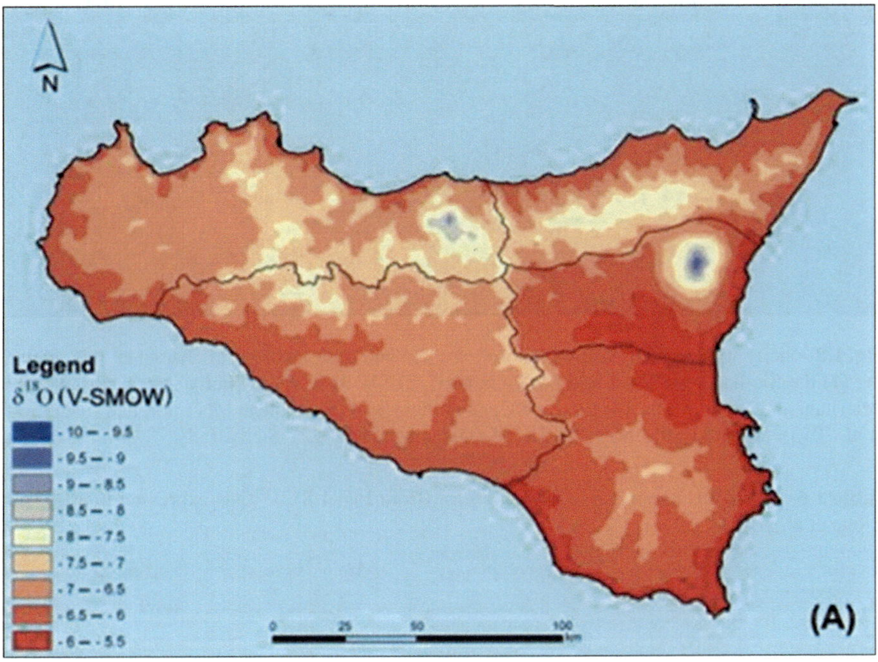

Fig. 1.7 Isotopic composition of $\delta^{18}O$ in rainwater in Sicily (modified from Liotta et al. 2013)

Each sector is assimilated to one or more hydrogeochemical basins:

1. Madonie, Palermo, Trapani, Marsala and Mazara del Vallo, Rocca Busambra, Castelvetrano and Campobello di Mazara;
2. Sicani, Piazza Armerina, Caltanissetta, Licata, Gela;
3. Nebrodi and Peloritani Mountains;
4. Mt Etna and Piana di Catania;
5. Hyblean Mountains.

Liotta et al. (2013) also determined the vertical gradient of ^{18}O useful to determine the feeding altitude of groundwater (Table 1.7).

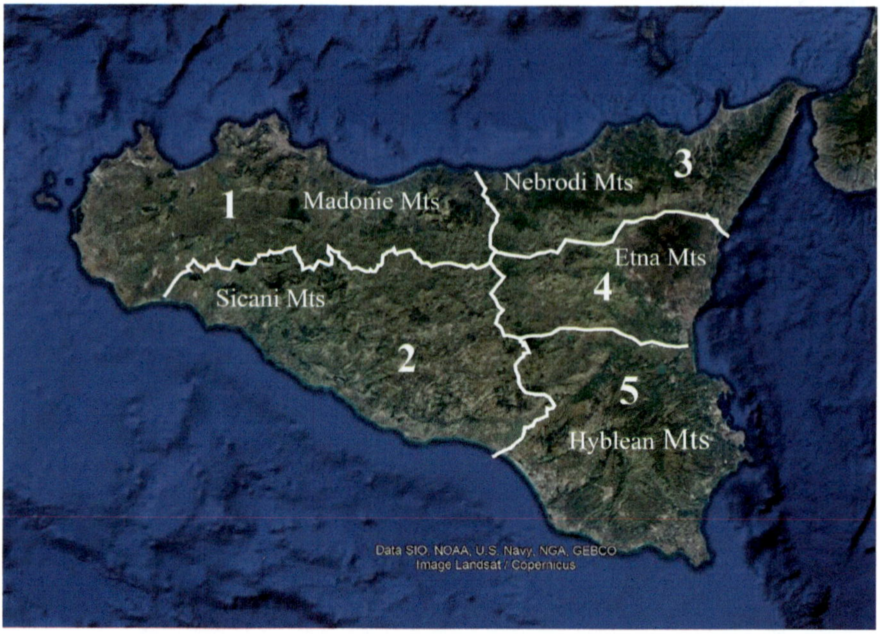

Fig. 1.8 Sicily was divided based on morphological and hydrogeological characteristics, including (1) the Sicilian west coast and Madonie Mts, (2) SW and central Sicily (Mt Sicani etc.), (3) Peloritani and Nebrodi Mts, (4) Etna and Catania Plain, and (5) Hyblean Mts (modified from Liotta et al. 2013). Google Earth Pro v 7.3.6.9345 (Dicember 29, 2022). Sicily, Italy

Table 1.6 WLMWL parameters for each sector (from Fig. 1.8). $\delta^{18}O$ is expressed in δ‰ (from Liotta et al. 2013)

Sector	Slope	Intercept	Min $\delta^{18}O$	Max $\delta^{18}O$
1	6.31	5.0	−8.1	−5.9
2	7.67	13.9	−7.5	−6.1
3	6.14	3.9	−8.2	−5.5
4	6.82	8.6	−10.5	−6.1
5	6.12	5.1	−7.3	−5.7

Table 1.7 Vertical gradient of each sector (from Fig. 1.8)

Sector	Vertical gradient
1	$\delta^{18}O‰ = -0.0015z-6.42$
2	$\delta^{18}O‰ = -0.0016z-6.18$
3	$\delta^{18}O‰ = -0.0013z-6.07$
4	$\delta^{18}O‰ = -0.0014z-5.69$
5	$\delta^{18}O‰ = -0.0016z-5.92$

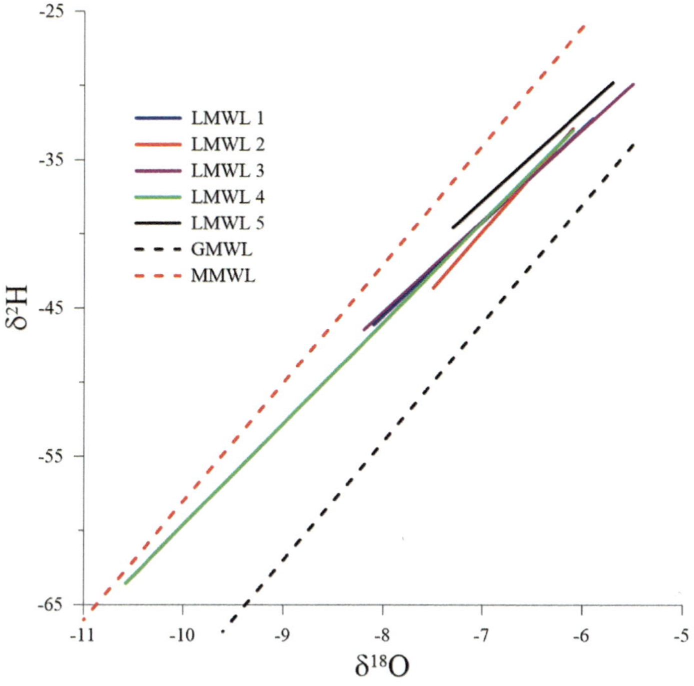

Fig. 1.9 Local meteoric water line for each sector (from Fig. 1.8, modified from Liotta et al. 2013). The Global Meteoric Water Line and Mediterranean Meteoric Water Line are also shown for comparison

In this study we use the specific LMWL and the $\delta^{18}O$ vertical gradient of each sector (and hydrogeological basin) taken from Liotta et al. (2013), except for those basins where specific, more exhaustive studies have been elaborated (Etna, Hyblean Mts, Rocca Busambra). The global reference line (Global Meteoric Water Line, $\delta^2H = 8\ \delta^{18}O + 10$; Craig 1961) and the Eastern Mediterranean Meteoric Water Line ($\delta^2H = 8\ \delta^{18}O + 22$; Gat and Carmi, 1970) are shown for comparison in the δ^2H versus $\delta^{18}O$ plot for each sector in Sicily (Fig. 1.9).

1.7 The Drinking Water Springs

In this study, we considered the sources reported in the Italian General Regulatory Plan for Aqueducts (PRGA) that were updated and revised in November 2010 by SOGESID. The PRGA contains all the resources restricted under Presidential Decree No. 1090 of 11 March 1968 and Law No. 36 of 5 January 1994 (Galli Law), and subsequent amendments and additions. The PRGA defines the current and future use of resources—in particular those for civil use. The surveyed resources consist of a detailed inventory of all the sources used by the Sicilian municipalities or managing bodies.

The aqueduct service in Sicily is organised in two segments:

(1) **Area aqueducts (*Acquedotti d'ambito*)**: made up of the nine *Ambiti Territoriali Ottimali* (ATO), coinciding with the administrative limits of the nine Sicilian Provinces, which manage supplies on a provincial or municipal scale;
(2) **supra-regional aqueducts (*Acquedotti di Sovrambito*)**: constituted by the large infrastructures for supplying and transporting water resources; characterised by a high degree of interconnection that allows the transfer of resources from areas with greater availability to areas where resources are lacking (Trapani and Palermo, inland areas; Messina, Agrigento, Caltanissetta, Enna) and the areas whose catchment area also falls within more than one ATO. The sources are divided into ten groups, one for each of the nine ATOs and one for the overlapping areas. The PRGA tables contain the most significant technical data of the resources.

Within the scope of the current study, the springs have been subdivided according to the hydrogeological basin they belong to, and to the reference water body. For each basin, a brief geological and hydrogeological description, the sampled springs (or for which it was possible to find chemical analyses) and the hydrogeochemical characterisation are reported. For the geochemical characterisation, in addition to the springs reported in the PRGA, a series of springs used by the various municipalities and local inhabitants for drinking were also used (Fig. 1.10), including several historical fountains in some towns, of which some photos are reported (Figs. 1.11, 1.12 and 1.13).

The data reported were taken from the existing bibliography, from surveys carried out directly by the authors, or from studies carried out for various purposes, supplemented by unpublished data.

Some areas have been thoroughly and extensively investigated and were represented by a myriad of articles on a wide variety of topics. In some areas, on the contrary, the bibliography is scarce. For this reason, some paragraphs are more newsworthy than others. In order to achieve a better, more complete geochemical characterisation of the catchment areas, additional data—not necessarily related to drinking water sources—were used.

Consulted studies:

- Istituto Nazionale di Geofisica e Vulcanologia (INGV), Sezione di Palermo (2005) Prima caratterizzazione e monitoraggio delle Acque Sotterranee finalizzata alla redazione del "Piano di Tutela delle Acque della Regione Sicilia". Per conto della Regione Siciliana—Commissario Delegato per l'Emergenza Rifiuti e Tutela delle Acque.
- Ministero dei Lavori Pubblici, Servizio Idrografico (1934) Le sorgenti italiane—elenco e descrizione. Publication n° 14 of the Service, vol. II, Sicily. Istituto Poligrafico dello Stato, Rome.

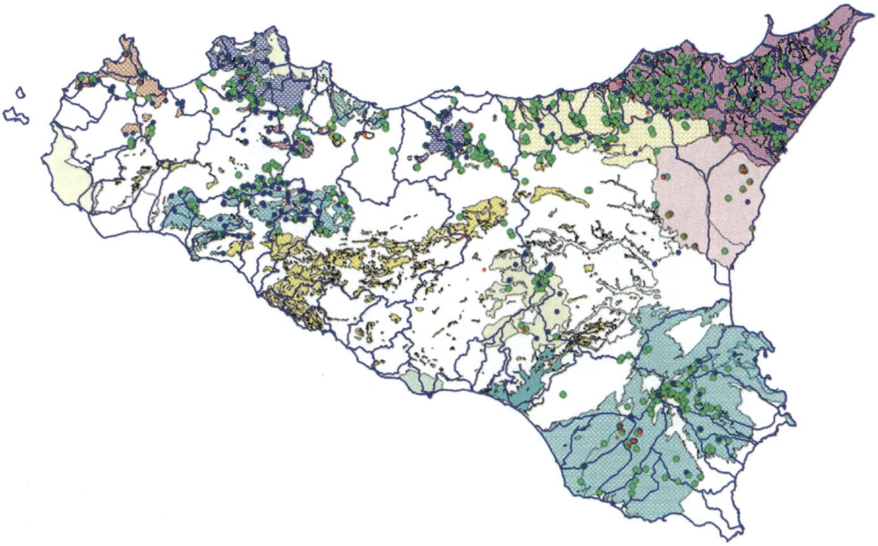

Fig. 1.10 Location of springs: green dots are PRGA springs, blue dots represent PTA springs and red dot show the location of the springs sampled in this study

Fig. 1.11 Abate Fountain (Alcara Li Fusi)

Fig. 1.12 Quattro Cannoli Fountain (Castelbuono)

Fig. 1.13 Trono Fountain (Raccuja)

- Scientific publications in various national and international journals listed as references at the end of each chapter.
- Cassa per il Mezzogiorno (1976–1977) Water Plan of Sicily.
- Cassa per il Mezzogiorno (1980) Studies and investigations related to the retrieval of underground water for the water supply of the North-Western I system of Sicily. Rip. Water Projects (Div. 3), Dagh Watson (unpublished study).
- Cassa per il Mezzogiorno (1982) Studies and investigations related to the finding of groundwater for the Water Supply of the I North-Western system of Sicily—Madonie. Rip. Water Projects (Div. 3), Dagh Watson (unpublished study).
- Cassa per il Mezzogiorno (1982) Hydrogeological and geophysical investigations for the retrieval of groundwater for the water supply of the system V central-eastern area of Sicily (Catanese). CMP S.p.A. (unpublished study).
- MINISTERO LL.PP. (1963) *Piano Regolatore Generale Acquedotti: Sicilia*
- Regione Siciliana (2012) Aggiornamento e Revisione del Piano regolatore generale degli Acquedotti. Relazione Generale del P.R.G.A. Convenzione di affidamento delle attività di supporto alla Regione Siciliana per l'attuazione degli interventi di interesse regionale ai sensi della L.R. 10/99. Sogesid, November 2010.

1.8 Sampling and Analytical Methods

The sampling, storage and analytical methods are summarised in Table 1.8. At each sampling point, the chemical and physical parameters were measured using portable field instruments (temperature, pH, electrical conductivity and redox potential). The samples taken were analysed in the laboratory, for the determination of the major anions (F, Cl, Br, NO_2, NO_3, SO_4, PO_4, Alkalinity) and cations (Na, K, Mg, Ca), and minor and trace elements (Li, Be, B, Al, V, Total Cr, Fe, Mn, Ni, Co, Cu, Zn, Se, As, Rb, Sr, Mo, Cd, Sb, Ba, Tl, Pb, Bi, U).

Sampling was carried out according to the indications given in EPA protocol n°441/2000 "A guide to the sampling and analysis of waters, wastewaters, soils and wastes" and in the APAT, IRSA-CNR protocol (29/2003) "Analytical methods for water—1030 Sampling methods". A total of 125 water samples were taken (springs and drainage tunnels).

With regard to the evaluation of analytical errors, the difference between the sum of the main cations and that of the anions, expressed in equivalents per litre, was always less than 3% for all the water samples considered.

Table 1.8 Outline of the sampling and analytical methods operated in this study

Parameters	Storage	Collected Volume (ml)	Sample conservation	Maximum storage time	Analitical method	Notes
Anions (F, Cl, Br, NO_2, SO_4)	HDPE plastic bottle, filtered with 0.45 μm filter	100	T ~ 4 °C stored in the dark	7 days	Ionic Chromatography (Dionex ICS-1100)	
Alkalinity					Titration with HCl, methyl orange and phenolphtalein	
Cations (Na, K, Ca, Mg)	HDPE plastic bottle, filtered with 0.45 μm filter	50	T ~ 4 °C stored in the dark	24 h	Ionic Chromatography (Dionex ICS-1100)	Stabilised at pH < 2 with drops of HNO_3 1%v/v suprapure
Minor and trace elements (Li, Be, b, Al, V, toal Cr, Fe, Mn, Ni, Co, Cu, Zn, Se, As, Rb, Sr, Mo, Cd, Sb, Ba, Tl, Pb, Bi, U)	HDPE plastic bottle, filtered with 0.45 μm filter	50	T ~ 4 °C V in the dark	7 days	ICP-MS (Agilent 7500ce) Standard In and Rn	Stabilised at pH < 2 with drops of HNO_3 1%v/v suprapure
Mercury (Hg)	Pyrex glass bottle, filtered with 0.45 μm filter	50	T ~ 4 °C stored in the dark	7 days	ICP-MS (Agilent 7500ce) Standard In and Rn	Stabilised at pH < 2 with drops of HNO_3 1%v/v suprapure
Cyanides/CN)	HDPE plastic bottle, filtered with 0.45 μm filter	100	T ~ 4 °C stored in the dark	7 days	Spectrometry (Shimatzu UV visible UV-1601)	Stabilised at pH < 12 with drops of NaOH

(continued)

Table 1.8 (continued)

Parameters	Storage	Collected Volume (ml)	Sample conservation	Maximum storage time	Analitical method	Notes
Esavalent Cromide	HDPE plastic bottle, filtered with 0.45 μm filter	50	T ~ 4 °C stored in the dark	24 h	Spectrometry (Shimatzu UV visible UV-1601)	Stabilised at pH < 2 with drops of HNO_3 1%v/v suprapure
Ammonium (NH_4)	HDPE plastic bottle, filtered with 0.45 μm filter	50	T ~ 4 °C stored in the dark	7 days	Spectrometry (Shimatzu UV visible UV-1601)	Stabilised at pH < 2 with drops of HNO_3 1%v/v suprapure
δ^2H, $\delta^{18}O$ of water	HDPE plastic bottle, unfiltered	50			Mass Spectrometry	

1.9 Data Processing

Thematic maps were drawn up to illustrate the concentration and distribution of the main parameters over the territory. The thematic maps produced are described below.

Figure 1.14 shows a map of Sicily where all the sources of the PRGA are located with the diameters of the circles proportional to the average flow rates. The sources have been grouped in six classes. Figure 1.14 highlights that the sources with the highest flow rates are located in the Etna area, in the Madonie Mountains and in the Monti di Palermo area.

In order to have an overview of the conductivity (and hence the salinity) of the analysed springs, the map in Fig. 1.15 was constructed. The highest values are related to the chalky-sulphur outcrop areas and the coastal areas that are influenced by seawater incursion.

1.10 Trace Elements

The presence of trace elements in spring waters for drinking purposes must be taken into account due to the toxicity of some if their concentrations exceed certain values (Fig. 1.16). Together with Human activity (anthropogenic pollution), some natural processes, such as volcanic activity, chemical alteration of rocks and mineralization (mineral solubility, precipitation/dissolution reactions, sorption/

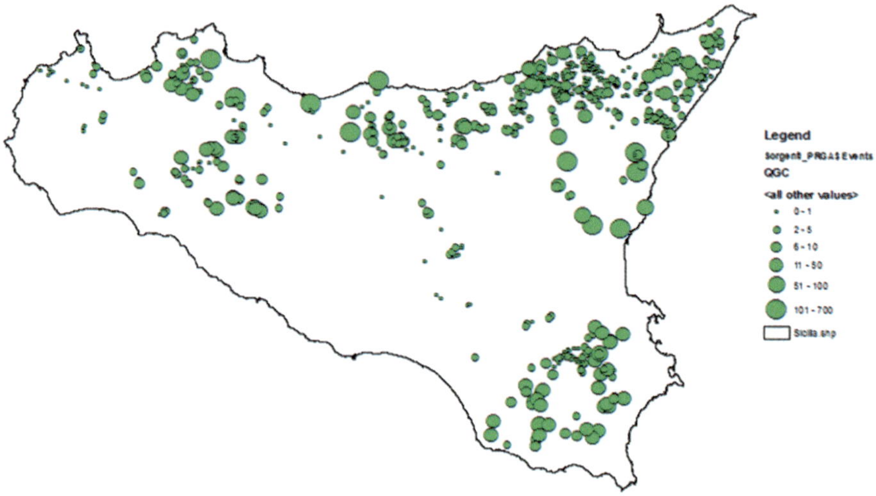

Fig. 1.14 Map of average flow rates of PRGA sources

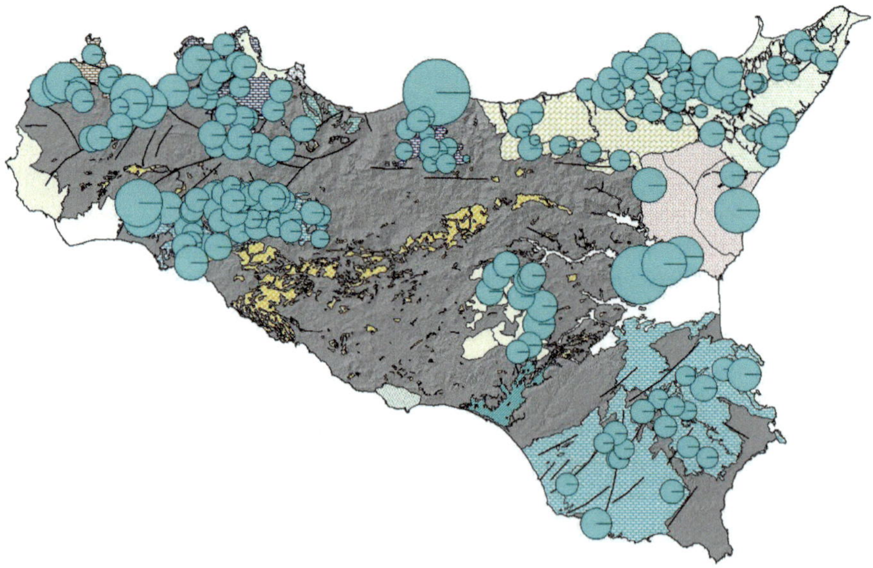

Fig. 1.15 Conductivity distribution map of the analysed sources

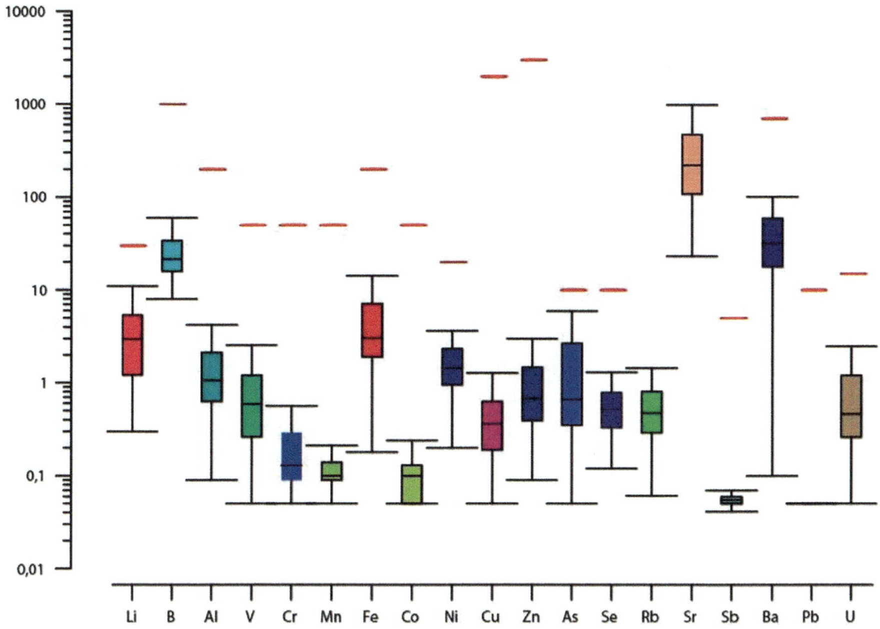

Fig. 1.16 Box-Whisker diagram of the trace elements examined for all the springs in this study. The red dash represents the legal limit for toxicity (D.Lgs. 30/2009)

desorption processes) can contribute to an increase in trace element contents up to harmful levels for human health. As can be seen from the diagram (Fig. 1.16), all the data considered so far are below the legal limits. The high concentrations of Sr and Ba are typical of interactions between water and carbonate rocks. The geochemical mobility of trace elements is strongly related to temperature, pH and redox condition of the water (Stumm and Morgan 1996). Explaining the highly complex geochemical mobility of trace elements falls outside the scope of this monograph (for more details see i.e. Stumm and Morgan 1996; AAVV 1995). The distribution of trace elements in Sicilian groundwater is shown in the map in Fig. 1.17 as pie charts for each site.

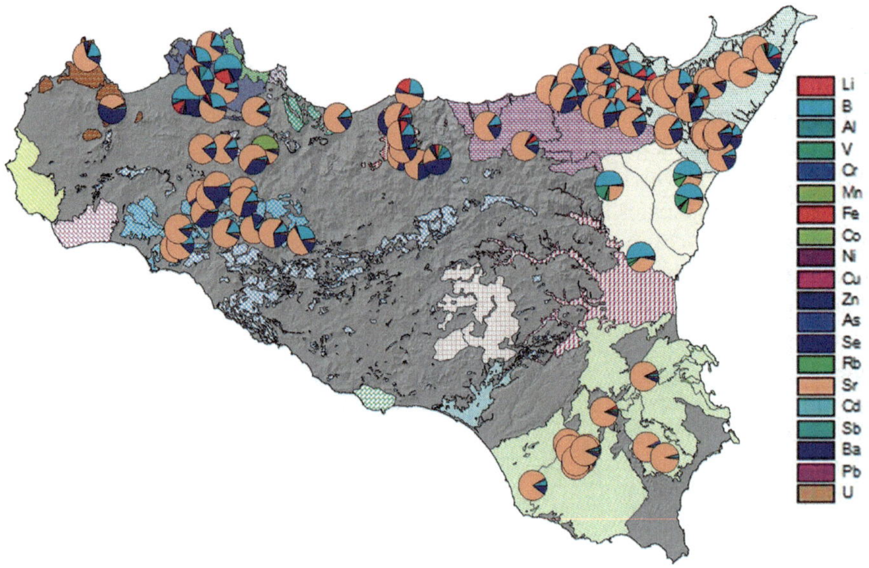

Fig. 1.17 Hydrogeological basins, distribution and relative proportions of trace elements in groundwaters of Sicily

Bibliography and Recommended Further Reading

AAVV (1995) Trace elements in natural water. In: Salbu B, Steinnes E (eds). CRC Press, p 302. ISBN 0-8493-6304-7

Cassa Per Il Mezzogiorno (1976) Water Sicily plan, study of groundwater in Sicily. CMP, Roma

Cassa per il Mezzogiorno (1980) Studies and investigations related to the retrieval of underground water for the water supply of the North-Western I system of Sicily. Rip. Water Projects (Div. 3), Dagh Watson (unpublished study)

Cassa per il Mezzogiorno (1982) Studies and investigations related to the finding of groundwater for the Water Supply of the I North-Western system of Sicily—Madonie. Rip. Water Projects (Div. 3), Dagh Watson (unpublished study)

Catalano R, D'Argenio B (1982) Guide to the geology of Western Sicily. Soc Geol It, Guide Geol Reg, 155 pp

Catalano R, Di Stefano P, Sulli A, Vitale FP (1996) Palaeogeographic and structural evolution of Sicily and adjacent seas. Natur Sicil S IV 19(3–4):143–187

Civita M (2008) The hydrogeological structure of the Italian territory: resources and problems. Quaderni Della Società Geologica Italiana 3:34

Craig H (1961) Isotopic variation in meteoric waters. Science 133:1702–1703

D'Alessandro W, Federico C, Longo M, Parello F (2004) Oxygen isotope composition of natural waters in the Mt Etna area. J Hydrol 296(1–4):282–299. https://doi.org/10.1016/j.jhydrol.2004.04.002

EPA (2000) A guide to the sampling and analysis of waters, wastewaters, soils and wastes. EPA Publication 441, 7th edn, Environment Protection Authority (Victoria)

Favara R, Gagliano Candela E (2005) General considerations on water availability in Sicily and problems related to the management of the resource. Istituto Nazionale di Geofisica e Vulcanologia, Sezione di Palermo. Priulla, Palermo, p 181

Gat JR, Carmi H (1970) Evolution of the isotopic composition of atmospheric waters in the Mediterranean Sea area. Geophys Res 75:3039–3040

Grassa F, Favara R, Valenza M (2006) Moisture source in the Hyblean Mountains region (southeastern Sicily, Italy): evidence from stable isotopes signature. Appl Geochem 21:2082–2095. https://doi.org/10.1016/j.apgeochem.2006.07.014

IAEA (1981) Stable isotope hydrology—deuterium and oxygen-18 in the water cycle. In: Gat, Gonfiantini (eds.). International Atomic Energy Agency. Vienna. STI/DOC/10/210. ISBN 92-0-145281-0

Liotta M, Grassa F, D'Alessandro W, Favara R, Gagliano Candela E, Pisciotta A, Scaletta C (2013) Isotopic composition of precipitation and groundwater in Sicily, Italy. Appl Geochem 34:199–206. https://doi.org/10.1016/j.apgeochem.2013.03.012

Liotta M, Brusca L, Grassa F, Inguaggiato S, Longo M, Madonia P (2006a) Geochemistry of rainfall at Strombolivolcano (Aeolian Islands): Isotopic composition and plume-rain interaction. Geochem Geophys Geosyst 7:Q07006. https://doi.org/10.1029/2006GC001288

Liotta M, Favara R, Valenza M (2006b) Isotopic composition of the precipitations in the central Mediterranean: origin marks and orographic precipitation effects. J Geoph Res 111:D19302. https://doi.org/10.1029/2005JD006818

Longinelli A, Selvo E (2003) Isotopic composition of precipitation in Italy: a first overall map. J Hydrol 270:75–88

Meinzer OE (1923) The occurence of groundwater in the United States with a discussion of principles. U.S. Geological Survey Water Supply Paper 489

Ministero dei Lavori Pubblici—Servizio Idrografico (1934) The Italian springs-list and description. Pubblicazione n° 14 del Servizio, vol. II, Sicilia. Istituto Poligrafico dello Stato, Roma

Istituto Nazionale di Geofisica e Vulcanologia (INGV), Sezione di Palermo (2005) First characterization and monitoring of groundwater aimed at the drafting of the "Water Protection Plan of the Sicily Region". On behalf of the Sicilian Region—Deputy Commissioner for the waste emergency and water protection

PAT-IRSA CNR (2003) Analytical methods for waters. APAT Manuali e Linee guida 29/03

Pecoraino G, Scalici L, Avellone G, Ceraulo L, Favara R, Gagliano Candela E, Provenzano MC, Scaletta C (2008) Distribution of volatile organic compounds in Sicilian groundwaters analysed by head space-solid phase micro extraction coupled with gas chromatography mass spectrometry (SPME/GC/MS). Water Res 42:3563–3577

Stumm W, Morgan JJ (1996) Aquatic chemistry, 3rd edn. Wiley, New York, p 1022

2 Mount Etna

Sabina Morici, Leonardo La Pica, Esterina Gagliano Candela, and Giovannella Pecoraino

2.1 Geological Framework

Mount Etna, with a surface area of approximately 1500 km^2, is a large strato-volcano with an alkaline basalt-Hawaiian composition (trachybasalt), located near the east coast of Sicily (latitude 37°30′–37°55′ N, longitude 14°47′–15°15′ E), bounded to the north by the Alcantara River valley, to the west by the Simeto River, to the south by the Catania Plain and to the east by the Ionian Sea (Fig. 2.1).

The origin of Etna volcano dates back to the Upper Miocene and is attributed to the rupture of the African plate margin during the collision with the European continental block (Barberi et al. 1974). On a large scale there is a system of NE-SW and NW–SE faults that separate the Maghrebid Apennine chain to the north, belonging to the Eurasian plate, from the Hyblean platform to the south, belonging to the African plate (Lentini 1982).

In more detail, within this broad regional context, Etna was formed at the intersection of three fault systems that cut through the thick continental crust:

- The Tindari-Giardini system oriented NNW-SSE (Carbone et al. 1987);
- The Comiso-Etna system oriented NE-SW (Lo Giudice et al. 1982; Bousquet et al. 1988);

S. Morici · L. La Pica · E. Gagliano Candela (Deceased) · G. Pecoraino (✉)
Istituto Nazionale di Geofisica e Vulcanologia, Sezione di Palermo, Via Ugo La Malfa 153, 90146 Palermo, Italy
e-mail: giovannella.pecoraino@ingv.it

S. Morici
e-mail: sabina.morici@ingv.it

L. La Pica
e-mail: leonardo.lapica@ingv.it

E. Gagliano Candela et al. (eds.), *Drinking Water in Sicily*,
https://doi.org/10.1007/978-3-031-60473-7_2

Fig. 2.1 Mount Etna. Image from Google Earth Pro v 7.3.6.9345 (Dicember 29, 2022). Sicily, Italy

- The Mt Kumeta-Alcantara system oriented E-W (Ghisetti and Vezzani 1984).

The volcanic complex developed to the NW on flyshoid sediments and to the SE on Pleistocene clays, both belonging to the Maghrebid Apennine chain and overlying the 10 km thick carbonate sequence (Fig. 2.2). Furthermore, the presence of evaporitic rocks lying above the sedimentary sequence is hypothesised under the lavas on the southern flank of Mount Etna (Lentini 1982).

2.2 Hydrogeology

Mount Etna is considered the most important groundwater reservoir in Sicily. The Etna massif represents a hydrogeological unit made up of volcanites that can reach a summed thickness of several hundred metres. The high permeability (10^{-5}–10^{-7} cm/s) of most volcanic products (Ferrara 1975a, b; Schilirò 1988) determines a high percentage of infiltration of meteoric precipitation (75%) and therefore the existence of water tables of significant potential that flow towards the peripheral areas of the volcano, supported by the impermeable sedimentary basement.

The structural geological setting of the impermeable sedimentary basement conditions the direction of the outflows because they are linked to the existence

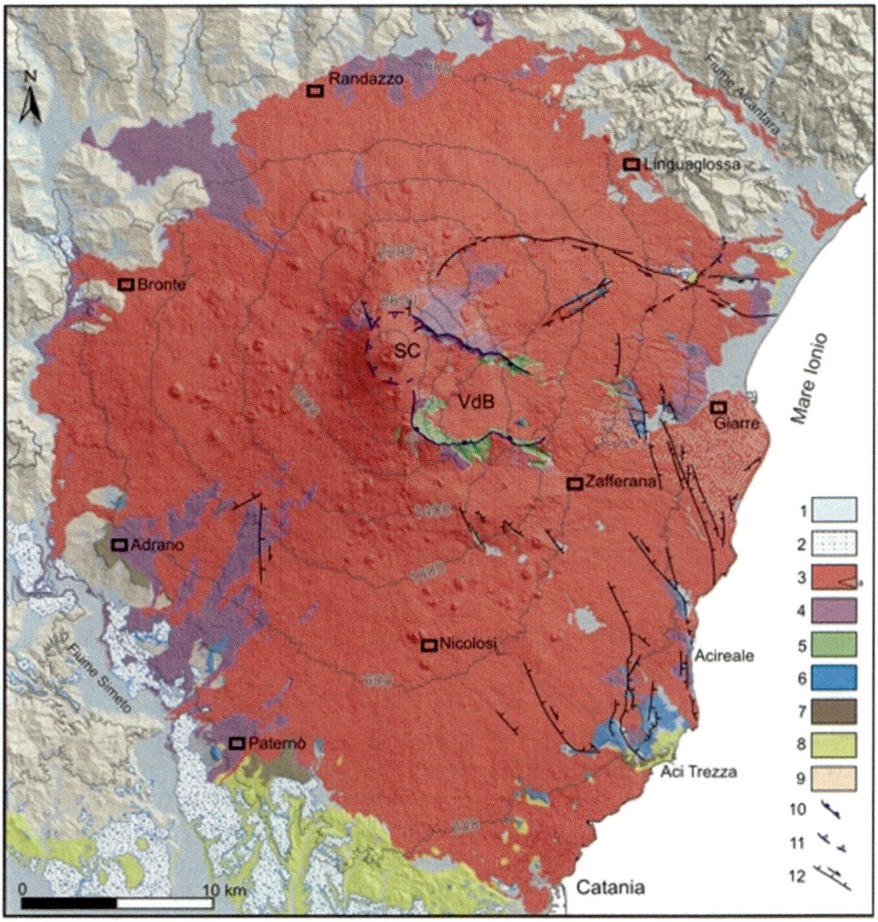

Fig. 2.2 Schematic geological map of Mount Etna (from Ruggieri 2022). *Legend* (1) recent and present alluvial deposits; (2) ancient alluvial deposits; volcanic products of the stratovulcano phase; (3) Mongibello volcano; (4) Elliptical volcano; (5) volcanic products of the Bove Valley Centres phase; (6) volcanic products of the Timpe phase; (7) volcanic products of the Basal Tholeiites phase; (8) Lower-Middle Pleistocene sediments; (9) sedimentary soils of the Apennine-Magrebid chain; (10) Bove Valley rim (VdB); (11) caldera rim; (12) direct faults, the arrow indicates the direction of movement. SC = summit craters

Table 2.1 Etna water bodies and their description

Water body	Hydrogeological complex	Mouton hydrogeological complex	Description	PTA Hydrogeological complex	Type of Aquifer
Etna East	VU Volcanites	Complex of pyroclastites, lavas and volcanites in general	Ancient volcanic soils and lavas (tuffs, sands and 'Chiancone' conglomerates); Recent lavas	Etna basic volcanite complex	Multilayer aquifer
Etna North	VU Volcanites	Complex of pyroclastites, lavas and volcanites in general	Etna and M. Mojo effusive products	Etna basic volcanite complex	Multilayer aquifer
Etna West	VU Volcanites	Complex of pyroclastites, lavas and volcanites in general	Ancient volcanic lavas and soils (tuffs, breccias); Recent basaltic lavas	Etna basic volcanite complex	Multilayer aquifer

of underground watersheds that delimit primary and secondary aquifer structures (Table 2.1).

The sedimentary basement beneath the volcanic building reaches an altitude of 1300 m a.s.l. a few kilometres NW of the volcano summit and dips towards the SE (Fig. 2.3), causing most of the water to flow and accumulate in the SE flank where there numerous springs and wells are located (Ogniben 1966). In accordance to the morphology of the sedimentary basement, as revealed by geophysical and geological data, three main hydrological basins can be identified, separated by three underground watersheds (Figs. 2.4, 2.5; Table 2.1): the first, to the north, feeds the Alcantara River; the second, to the W-SW of the volcanic edifice, feeds the Simeto River; finally, the largest basin, to the east of Etna, is a tributary of the Ionian Sea.

In the north basin, below the volcanic soils that represent the aquifer, we find the sedimentary soils of the Apennine-Maghrebide chain, which form the impermeable substratum delimiting the water table downwards (Fig. 2.3). The morphology of this substratum, which characterises the groundwater flow in this sector, is the result of the interaction between the paleo-rifts of the Alcantara watershed and the lava flows that have interrupted the river's course several times. The partial reconstruction of the palaeo-hydrography (based on geophysical surveys, well stratigraphy, hydrography existing today on the sedimentary soils of the right bank of the Alcantara River, the visible course of the lava flows and the geological survey of the sedimentary outcrops among the lava soils), has identified four palaeo-watersheds that were formed following the shifting of the Alcantara riverbed due to the barrages caused by the lava flows of both the Mt. Moio volcanic complex and subsequently Etna.

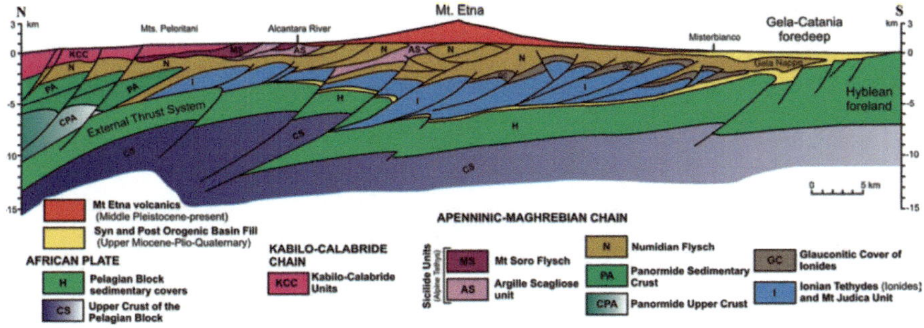

Fig. 2.3 Schematic N-S geologic cross section of Mt Etna Area (from Branca et al. 2011)

In the western basin, based on a geoelectrical survey, the course of the impermeable substrate was reconstructed up to an altitude of 1300 m a.s.l. and three paleo valleys were identified below the volcanic products. Two of these, defined as "northern paleo valleys" and "southern paleo valleys", fan out from Etna's summit area towards the base of the volcano. The third paleo-wall has been identified parallel to the course of the Simeto River and to the east of it; it is the ancient course of the Simeto River that has moved, due to the continuous lava contributions, towards the more erodible sedimentary areas. This paleo-wall begins at Ponte Cantera and ends just south of Ponte Maccarrone, where it joins the current course of the river that drains all the water from the Etnean slopes. The materials that make up the filling of the paleo-valleys are both lava and alluvial products; the presence of dams, consisting of quartzarenitic banks or more compact flows, causes lateral overflows of the water table towards the current Simeto riverbed, giving rise to a series of aligned springs with very similar chemical-physical and organoleptic characteristics. These springs identify the western edge of this paleo-valley that is 100 to 400 m from the left bank of the current Simeto riverbed.

The soils covering these paleo-valleys are formed by volcanic flow deposits with more or less thick layers of materials in between, such as alluvium, pyroclastites with varying degrees of alteration, products resulting from the degradation of the surface areas of the flow, and paleosoils. The type of permeability that characterises the flows is essentially linked to porosity through cracking, while porosity through interstices is low, if not nil. Horizons consisting of paleosoils and partly altered pyroclastites have little or no permeability, are interspersed with the flows and give rise to numerous suspended aquifers, with resurgences at higher altitudes than those of the base level, which are hardly connected to the deep aquifer. The deep aquifer, which is bound at the base by the two northern and southern paleovallates, is considered to be free in the broadest sense, with local artesian phenomena.

In the foothills between Adrano and Paternò, several aquifers can be recognised whose outflows reach the margins of the lava expansions, giving rise, on contact with the sedimentary soils or with the basal volcanites, to groups of springs that were once of considerable high capacity. Today, the yield of these springs is strongly reduced due to the large number of wells upstream, which have also led

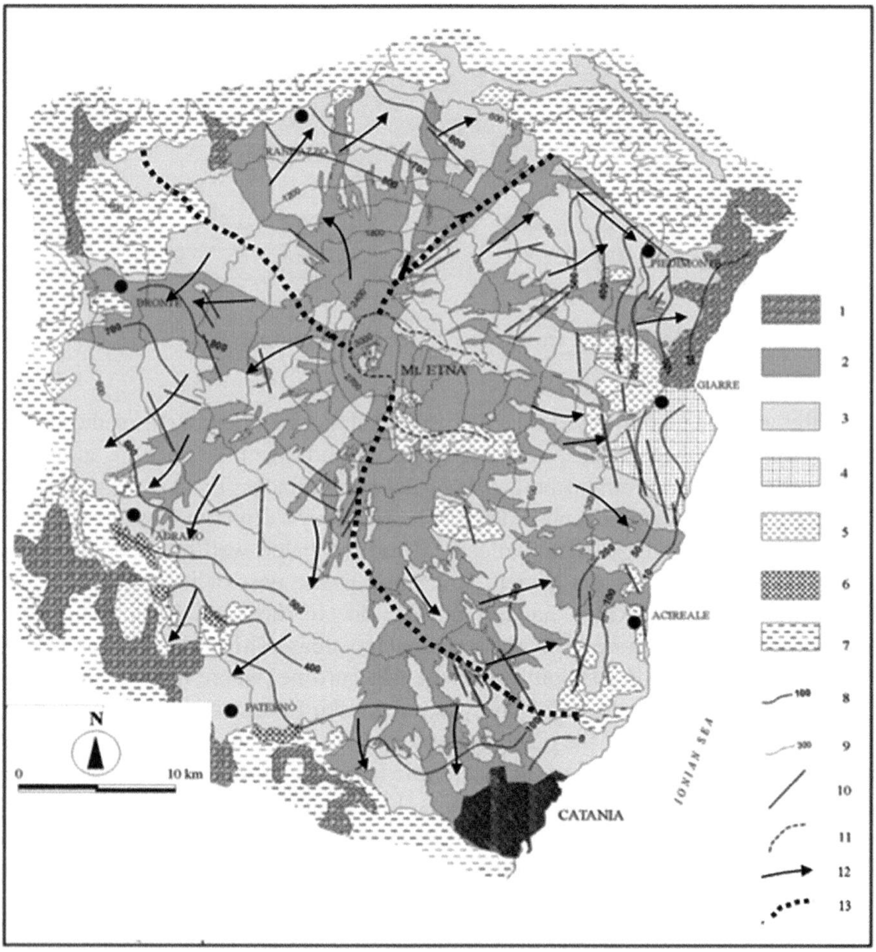

Fig. 2.4 Schematic hydrogeological map of the volcanic massif. (1) Alluvial deposits; (2) Historic lava flows and pyroclastic deposits; (3) Recent lava flows and volcanoclastic deposits; (4) Sandy alluvial conglomerates ("Chiancone"); (5) Ancient volcanic products (lavas, tephra, mudflows and tuffaceous sediments); (6) Basal volcanic products; (7) Sedimentary basement; (8) Water table contour lines (m a.s.l.); (9) Topographic contour lines (m asl); (10) Main faults; (11) Caldera rims; (12) main lines of underground flow; (13) geological watershed (modified from Ferrara 1991)

to a considerable lowering of the water table throughout the foothills. In these two sectors, the distribution of extraction works is characterised by a strong concentration (up to more than 6 wells per km^2) at lower altitudes, although there are also extraction points at high altitudes (up to 900–1100 m a.s.l.) whose productivity is generally high, involving the base aquifer located at depths of several hundred metres.

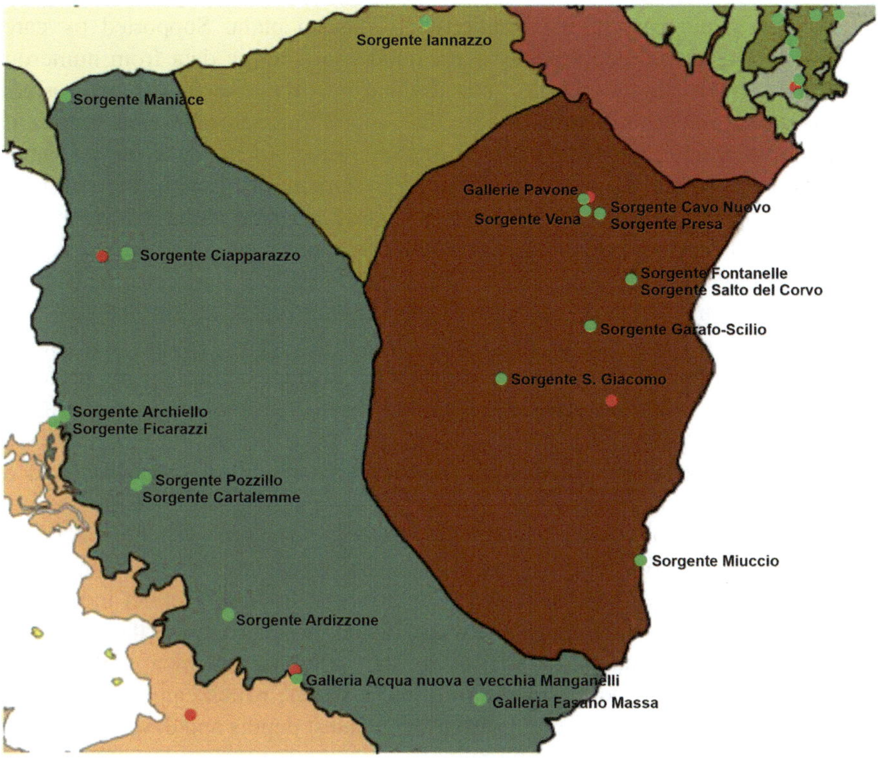

Fig. 2.5 Map of Mt Etna hydrological basin. Location of spring is shown (green dots represent PRGA spring, red dots spring sampled for this work). Colours identify water bodies: Grey Etna West; brown Etna East; yellowish Etna North

The eastern Etna basin extends from the summit areas to the Ionian coast and is bordered to the north by the valley of the Fogliarino-Minissale torrent (Piedimonte Etneo area) and to the south by the Misterbianco-Catania junction. As with the other slopes, the impermeable substratum is formed by the sedimentary soils on which the volcanic edifice was built, while the aquifer is mainly composed of volcanites. In the areas where the volcanites are interspersed with tuff and breccia horizons, two separate aquifers are formed that share a common supply from the summit areas of the volcano. Another aquifer is the *Chiancone* unit (Fig. 2.4) that emerges in the central-eastern area of this slope. At the western edge of the outcrop area and even further upstream, this deposit interdigitates with the volcanites, creating a connection with the volcanic aquifer.

Another aquifer, also connected with the volcanites, is represented by the recent alluvial deposits of the Fiumefreddo-Mascali plain. Supported by careful structural analysis, geophysical studies and stratigraphic data from numerous wells and tunnels, this slope has been divided into three sectors (north-eastern, central-eastern and south-eastern; Figs. 2.4 and 2.5), each of which represents an independent aquifer structure. Each of these three structures is further subdivided into smaller drainage areas separated by structural highs in the substrate. This basin is the most complex from a hydrogeological point of view; both in the foothills and at higher altitudes there are numerous deep and suspended aquifers. Many of these aquifers are adequately exploited and contribute to the supply of drinking water to the city of Catania and other towns. The east- and south-east-facing outflows are directed towards the Ionian coastline that extends from Fiumefreddo to Catania, forming several distinct aquifers with preferential drainage axes in correspondence with the most depressed parts of the minor aquifer structures. The main drainage routes are located along the Linguaglossa-Piedimonte-Fiumefreddo, Val Calanna-Zafferana-Giarre, Trecastagni-Aci and S. Antonio-Acireale links (Fig. 2.4).

Along the aforementioned alignments, tectonic structures can be recognised that are oriented according to regional systems that regulate the directions of the most consistent underground flows in relation to the fracturing effect induced by the dislocations. This is evidenced by the higher productivity of the wells that fall along these alignments, as well as by the particular chemical and physical characteristics of the water extracted from these works, which can be traced back to the gaseous contributions coming from the greatest depths and developing along the most important tectonic discontinuities.

Tritium data from meteoric and groundwater circulating in the Etnean aquifers suggest short residence times (approximately one year), except for the SW basin, where the deepest circuits are associated with residence times of approximately 50 years. The sedimentary basement below the volcanic edifice causes groundwater to circulate at shallow depths and isolates from thermalisation. In the SW sector, and in particular in the Paternò area, hypersaline fluids that can be traced back to "brine" rise up, mix with the circulating groundwater and display characteristics typical of hydrocarbon reservoirs, being rich in CO_2 and CH_4.

The high permeability for cracking of lavas confers a high to very high vulnerability in relation to the type of volcanic product hosting the aquifer.

In the conglomeratic-sandy deposits, originated by the dismantling and erosion of old volcanic apparatuses, the vulnerability has been evaluated as medium, always considering the lithological variability and the depth of the water table from the ground level.

The free water table contained in the alluvial deposits in the marginal zones of the volcanic edifice, generally located at a depth of a few metres, has a high degree of vulnerability. The flyschoid alternations, with scarce water circulation and limited to the fractured arenaceous banks, and the clay units, with practically no groundwater circulation, which constitute the base on which the volcanites rest, have a low or zero degree of vulnerability.

The likelihood that water resources may be damaged by pollution is therefore high, given the lithostructural characteristics of the aquifer, the method of underground water circulation, the depth from ground level. Considering also that most of the catchment works used to supply water to many of the towns on the Ionian coast are located at the lowest, highly anthropized altitudes. The risk of pollution is much higher in the foothills, where the sources of potential pollution are represented by the large number of urban centres and areas of recent building expansion with associated infrastructure. As altitude increases, the density of settlements and productive activities tends to progressively decrease, until they disappear completely above 800–1000 m a.s.l.. At the same time, the depth of the water table with respect to the ground level tends to increase due to the rapid accentuation of the relief, so that, in general, the risk conditions decrease.

2.3 Hydrogeochemical Characterisation

Previous studies on the geochemistry of groundwater from Mount Etna (Parello et al. 2001; Aiuppa et al. 2003) have highlighted the factors controlling the evolution of the chemical composition of groundwater. The main mechanism for the acquisition of ionic species in solution is governed by the dissolution in meteoric recharge waters of large quantities of acid gases, mainly of magmatic origin, and by the direct groundwater-gas interaction. The magmatic gases dissolved in the Etna aquifers are mainly composed of 'deep' CO_2 (which is transformed into water as the HCO_3^- ion) and smaller quantities of SO_2, HCl and HF (which are transformed into water as $SO_4^=$, Cl^- and F^-). The acidity thus acquired by the groundwater is buffered by the dissolution of the minerals forming the aquifer rocks. A contribution of SO_2 by the interaction between meteoric waters and the gaseous plume continuously emitted from Etna's summit craters has been recently demonstrated (Liotta et al. 2016).

The chemistry of the groundwater is also controlled by some secondary processes such as mixing with small amounts of seawater, or saline waters with a composition similar to seawater coming from the sedimentary basement below the volcanic aquifer (Chiancone Unit; Parello et al. 2001), and by the infiltration of Ca-rich water from the Nebrodi Mts through a system of paleo riverbed (Aureli and Musarra 1975). Moreover, also anthropogenic causes, first of all the overexploitation of aquifers, change the composition of groundwater (D'Alessandro et al. 2010).

In Etna basin, 18 springs for drinkable water are known (Table 2.2). Among them, nine have been sampled for this book, four were taken from INGV (2005) (Table 2.2).

Water springs are characterised by $10 < T < 17$ °C, EC range from 270 to 1500 mS/cm and a pH from 6.5 to 8 (Table 2.2).

Most of the drinking water sources in the Etna hydrogeological basin can be classified, according to the Langelier-Ludwig diagram (Fig. 2.6), as bicarbonate-alkaline-earthy, bicarbonate-alkaline and chlorinated-sulphate-alkaline.

Table 2.2 Chemical and isotopic composition of the sampled points

	UTM-E ED 50	UTM-N ED 50	T°C	EC	pH	Eh	Na	K	Ca	Mg	Cl	NO_3	SO_4	HCO_3	δ^2H	$\delta^{18}O$	WB
Miuccio	515,324	4,163,030					1.20	0.12	3.47	5.16	1.48	0.13	1.1	7.2			EE
Presa	512,880	4,182,710	12.3		8.00	60	1.19	0.11	0.28	0.35	0.46	0.01	0.31	1.05			EE
Salto del Corvo	514,610	4,179,040					1.24	0.03	0.87	7.75	1.77	0.32	0.37	7.4			EE
Vena	512,086	4,182,940	10.2		7.86	123	0.90	0.05	0.71	0.43	0.37	0.00	0.52	1.3		−6.09	EE
Pavone	512,050	4,183,680	13.2	616	7.3	185	3.98	0.36	0.76	2.66	1.17	0.01	2.01	4.75	−44	−7.83	EE
Garaffo e Scilio	512,574	4,176,115	13.3	273	7.88	110	1.68	0.21	0.48	0.63	0.78	0.00	0.71	1.44	−38	−7.05	EE
Iannazzo	503,120	4,193,600	12.4		6.71	134	2.54	0.3	2.58	3.00	0.9	0.18	1.01	6.3		−8.47	EE
Acqua nuova Manganelli	495,864	4,156,697	16.8	1528	6.76	169.6	5.88	0.85	4.36	11.21	2.92	0.38	1.51	17.00	−43	−7.46	EN
Ciapparazzo	484,880	4,180,332	11.5	1140	7.57	150.4	6.83	0.53	1.15	5.68	2.76	0.01	4.55	7.1	−43	−8.68	EW
Cartalemme 1	486,850	4,167,640	15.3		7.49	40	9.00	0.73	1.43	11.36	3.27	0.1	1.29	17.51		−8.81	EW
Cartalemme 2	486,850	4,167,640	14.1		8.59	288	9.17	0.82	1.37	10.21	2.6	0.05	1.58	17.1		−8.97	EW
Pozzillo	487,380	4,167,640	14.5		7.5	110	8.43	0.65	1.21	10.95	2.6	0.01	0.93	17.51		−6.80	EW
Ardizzone	492,075	4,159,930	15.3		6.57	367	5.02	0.41	2.52	5.96	1.59	0.07	0.34	11.8		−8.50	EW

Analytical data are expressed as meq/L, EC in μS/cm, Eh in mV and δ^2H and $\delta^{18}O$ in ‰ versus SMOW. WB is for water body. EE is Etna East, EN is Etna North and EW is Etna West

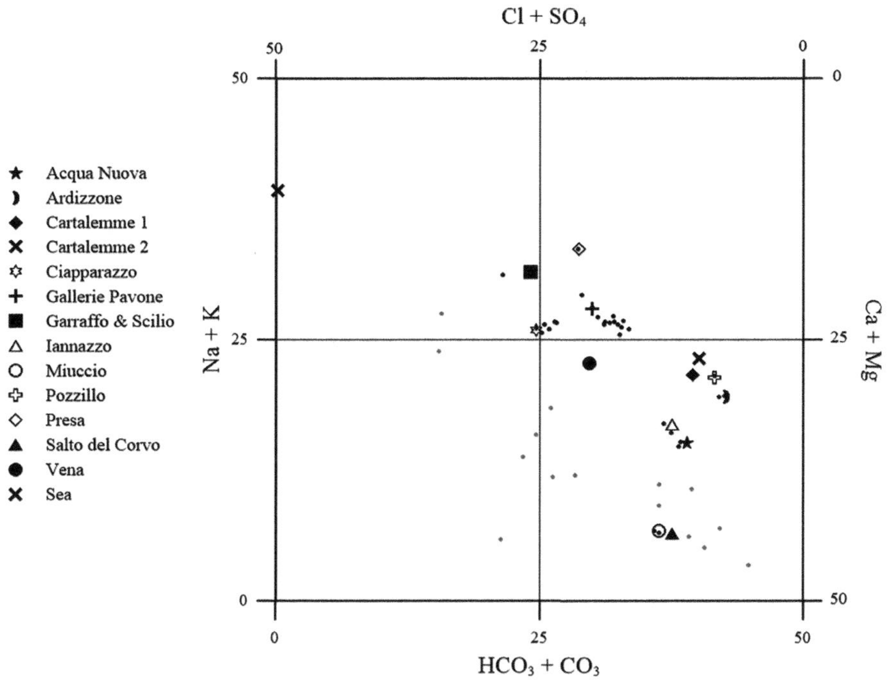

Fig. 2.6 Langelier-Ludwig diagram. The diagram shows the springs examined. Most of the springs fall within the bicarbonate-alkaline-earthy and bicarbonate-alkaline quadrants. The springs "Ciapparazzo" and "Garraffo & Scilio" fall between the quadrants "bicarbonate-alkaline" and "chlorinated-sulphate-alkaline". Literature data (wells and drainage tunnels) are shown for comparison (INGV 2005)

Due to the processes that groundwater undergoes during its flow (described above), HCO_3 is the dominant anion (Fig. 2.7) (D'Alessandro et al. 2010). Two alignments branch off from the HCO_3 corner, one towards the SO_4 corner representing waters (i) circulating in aquifers where the lithotypes present sulphur minerals derived from the alteration of old eruptive centres (Chiancone unit) (Parello et al. 2001), and (ii) fed by rain waters interacted with the Etnean plume (Liotta et al. 2016). The other alignment is towards the corner of the Cl and is represented by waters mildly mixed with brine with a Na/Cl ratio similar to seawater.

Among the cations, the dominant species are Na + K and are linked to interactions with waters of marine origin. Waters enriched in Ca (and often SO_4) flow in high S content rock as the Chiancone Unit.

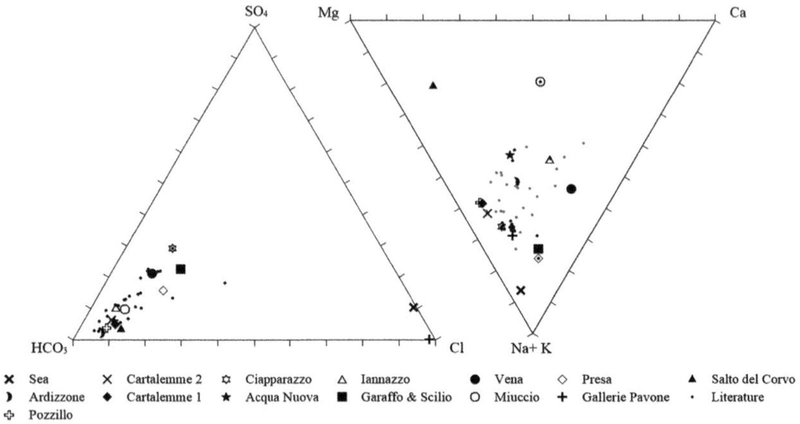

Fig. 2.7 Anions and cations ternary diagrams for Mount Etna catchment area. Springs that fall near the Na + K corner are those that are affected by mixing with "brine". The most Ca-rich springs are often enriched in anthropogenic sulphate or through interaction with reservoir rock. Enrichment in HCO_3 through interaction between groundwater and magmatic CO_2. Literature data are from Parello et al. (2001)

2.4 Isotopic Characteristics of the Catchment Area

The drinking water sampled in the basin for this study aligns very close to the Gat and Carmi (1970) meteoric water line for the Mediterranean Sea (Fig. 2.8).

The LMWL for Etna basin was determined by Parello et al. (2001) and by Liotta et al. (2013) as sector 4 in Table 1.6. Parello's LMWL ($\delta^2H = 8\delta^{18}O + 14$) was determined specifically for Mt Etna using 15 rain gauges during the 1997–1998 observation period. Liotta's LMWL ($\delta^2H = 6.75\delta^{18}O + 8.2$) considers a larger area of rainfall including part of the Catania Plain, 8 rain gauges during the 2004–2005 observation period.

Water data from samples and literature fall close to both LMWLs, confirming the predominant meteoric origin of these waters. According to Parello et al. (2001), based on the vertical isotopic gradient calculated in the area (0.27δ‰/100 m), Etnean groundwaters are fed from different altitudes depending on the water body to which they belong. In particular, the SW sector (corresponding to the Etna West water body) has an average feeding altitude of 930 ± 320 m a.s.l., those belonging to the eastern sector (corresponding to the Etna East water body) from altitudes of 330 ± 240 m a.s.l., and those belonging to the northern sector (corresponding to the Etna North water body) show feeding altitudes of 810 ± 220 m a.s.l. (Fig. 2.9).

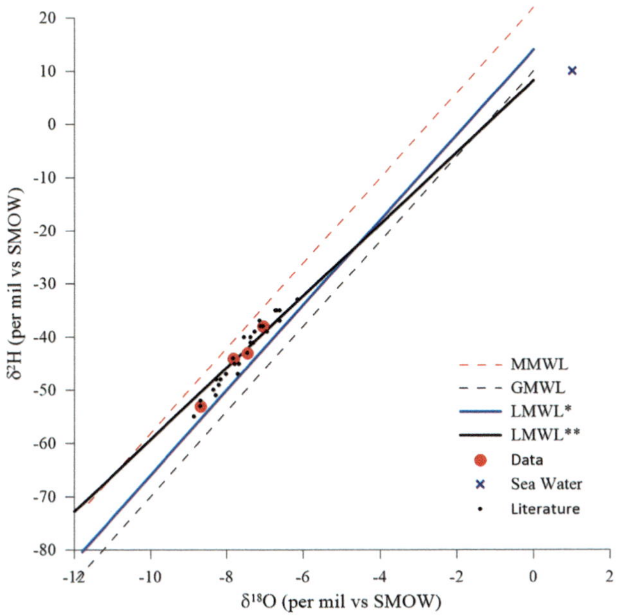

Fig. 2.8 δ^2H -δ^{18}O (in ‰ vs SMOW) diagram of spring waters. The LMWL* is from Parello et al. (2001), LMWL** is from Liotta et al. (2013). The GMWL (Craig 1961) and MMWL (Gat and Carmi 1970) and literature data are shown for comparison

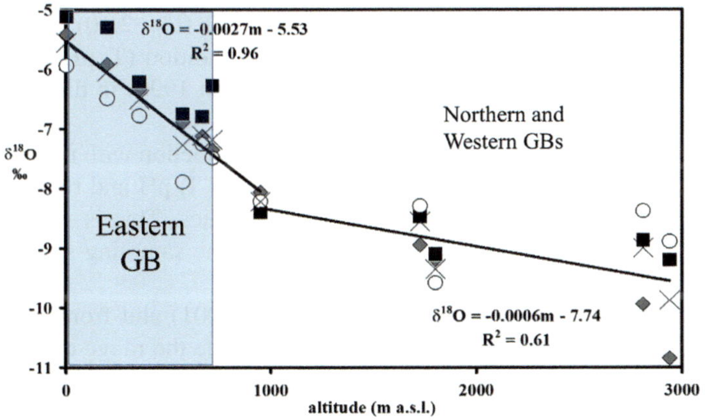

Fig. 2.9 Relationship between δ^{18}O and altitude. GB is groundwater basin. Data from D'Alessandro et al. (2004)

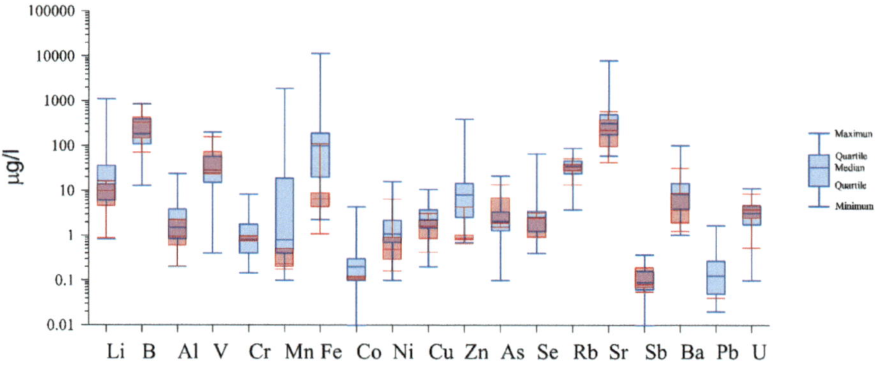

Fig. 2.10 Box-whisker plot of trace element at M. Etna basin. Red box spring data (this work). Ligh blu box data from Parello et al. (2001) and INGV (2005)

Most of Etna's groundwaters show low seasonal variability in δ^2H and $\delta^{18}O$ values, confirming a good isotopic homogenization in the aquifer supply circuits, which is probably due to the high fracture permeability of rocks, allowing easy mixing of water originating from different altitudes.

2.5 Trace Elements

In the Etna hydrogeological basin, trace elements show a wide range in concentrations varying from ppb (µg/L) to ppm (mg/L) levels (Fig. 2.10). This behaviour strongly depends on the water chemical-physical condition (T, pH, Eh) and gas–water-rock interaction variability (Stumm and Morgan 1996) in different areas of basin (Parello et al. 2001) (Table 2.3).

In the specific case of Etna groundwaters, their interaction with magmatic gases, linked to volcanic activity, cause a strong variation in T, pH and redox conditions and, hence, in the gas–water-rock interaction dynamics. This is why there is so much variation in each individual element at the same sampling site (see Parello et al. (2001) for additional details).

Figure 2.10 compares data from Parello et al. (2001) and from INGV (2005) with data collected in this study. Spring data fall inside the range of concentration of bibliographic data and always below the Italian legal limit (D.Lgs. 30/2009).

Table 2.3 Trace element concentrations (μg/L) in sampled waters

	Li	B	Al	V	Cr	Mn	Fe	Co	Ni	Cu	Zn	As	Se	Rb	**Sr**	Sb	Ba	Pb	U	WB
Manganelli	16.00	811	bdl	71.9	0.96	0.25	6.46	0.12	6.43	1.62	1.03	13.5	2.53	51.7	574	bdl	31.5	bdl	8.45	EN
Garraffo e Scilio	0.87	70.5	2.20	22.8	bdl	bdl	1.08	bdl	0.16	0.42	0.81	1.85	0.92	13.5	42.2	0.07	1.25	bdl	0.521	EE
Vena	1.40	nd	1.30	6.9	0.7	0.4	11.00	0.06	nd	0.4	35.7	0.3	nd	3.8	208	0.02	5.8	0.02	0.5	EE
Pozzillo	9.60	320	0.20	28.0	0.8	0.2	109	0.1	0.9	3.1	4.3	1.9	3.2	38.6	378	0.09	3.7	0.04	4.7	EW
Ciapparazzo	13.7	424	1.30	155	0.73	0.49	8.72	bdl	0.48	2.17	0.85	6.91	2.42	33.0	219	0.19	8.82	bdl	3.76	EW

Bibliography and Recommended Further Reading

Aiuppa A, Bellomo S, Brusca L, D'Alessandro W, Federico C (2003) Natural and anthropogenic factors affecting groundwater quality of an active volcano (Mt. Etna, Italy). Appl Geochem 18:863–882

Aureli A, Musarra A (1975) Hydrogeology of the Alcantara River Basin (Sicily). In: Proceedings 3rd international conference on underground waters, Palermo 1–3 Nov 1975, pp 157–216

Azzaro R, Branca S, Gwinner K, Coltelli M (2012) The volcano-tectonic map of Etna volcano, 1:100.000 scale: an integrated approach based on a morphotectonic analysis from high-resolution DEM constrained by geologic, active faulting and seismotectonic data. Ital J Geosci 131(1):153–170

Barberi F, Civetta L, Gasperini P, Innocenti F, Scandone R, Villari L (1974) Evolution of a section of the Africa-Europe plate boundary: paleomagnetic and volcanologic evidence from Sicily. Earth Planet Sci Lett 22:123–132

Bousquet JC, Lanzafame G, Paquin C (1988) Tectonic stresses and volcanism: in-situ stress measurements and neotectonic investigations in the Etnean area (Italy). Tectonophysics 149:219–231

Branca S, Coltelli M, Groppelli G, Lentini F (2011) Geological map of Etna volcano, 1:50,000 scale. Ital J Geosci 130(3):265–291. https://doi.org/10.3301/IJG.2011.15

Carbone S, Grasso M, Lentini F (1987) Geological features of the Hyblaean plateau (SE Sicily). Presentation of geological maps of southeastern Sicily. Mem Soc Geol Ital 38:127–135

Cassa per il Mezzogiorno (1983) Hydrogeological and geophysical investigations for the retrieval of groundwater for water supply of the V system, central eastern area of Sicily (Catanese). Prog Spec N 30. Unpublished report

C.N.R. (1979) Geological map of Mount Etna, scale 1:50,000. Artistic Cartographic Lithography, Firenze

Craig H (1961) Isotopic variation in meteoric waters. Science 133:1702–1703

D'Alessandro W, Federico C, Longo M, Parello F (2004) Oxygen isotope composition of natural waters in the Mt. Etna area. J Hydrol 296:282–299

D'Alessandro W, Bellomo S, Bonfanti P, Brusca L, Longo M (2010) Salinity variations in the water resources fed by the Etnean volcanic aquifers (Sicily, Italy): natural vs. anthropogenic causes. Environ Monit Assess. https://doi.org/10.1007/s10661-010-1397-4

Ferrara V (1975a) Hydrogeology of the eastern slope of Etna. In: Proceedings 3rd international conference on underground waters, Palermo 1–3 Nov 1975, pp 91–144

Ferrara V (1975b) Methods of groundwater capture and exploitation in volcanic soils of Etna. In: Proceedings 3rd international conference on underground waters, Palermo 1–3 Nov 1975, pp 474–481

Ferrara V (1994) Map of pollution vulnerability of etna volcanic aquifer, scale 1:50,000. C.N.R. National Group for the Defense of Hydrogeological Disasters, Pubbl N 266

Ferrara V, Pappalardo S (1991) The hydrogeological map of the Etna volcanic massif as useful tool for groundwater resource management. Ital J Eng Geol Environ Special Issue 1 (2008)

Gat JR, Carmi H (1970) Evolution of the isotopic composition of atmospheric waters in the Mediterranean Sea area. Geophys Res 75:3039–3040

Ghisetti F, Vezzani L (1984) Thin-skinned deformations of the Western Sicily thrust belt and relationships with crustal shortening: mesostructural data on the Mt. Kumeta-Alcantara fault zone and related structures. Boll Soc Geol It 103:129–157

Istituto Nazionale di Geofisica e Vulcanologia (INGV), Sezione di Palermo (2005) First characterization and monitoring of Groundwater aimed at the drafting of the "Water Protection Plan of the Region of Sicily". On behalf of the Sicilian Region - Delegate Commissioner for the Waste Emergency and Water Protection

Lentini F (1982) The geology of the Mt. Etna basement. Mem Soc Geol It 23:7–25

Liotta M, Grassa F, D'Alessandro W, Favara R, Gagliano Candela E, Pisciotta A, Scaletta C (2013) Isotopic composition of precipitation and groundwater in Sicily, Italy. Appl Geoch 34:199–206

Liotta M, D'Alcssandro W, Bellomo S, Brusca L (2016) Volcanic plume fingerprint in the groundwater of a persistently degassing basaltic volcano: Mt. Etna. Chem Geol 433:68–80
Lo Giudice E, Rasà R (1986) The role of the NNW structural trend in the recent geodynamic evolution of north-eastern Sicily and its volcanic implications in the etnean area. J Geodyn 25:309–330
Lo Giudice E, Patané G, Rasà R, Romano R (1982) The structural framework of Mount Etna. Mem Soc Geol It 23:125–158
Monaco C, Petronio L, Romanelli M (1995) Extensional tectonics in the eastern sector of Mount Etna (Sicily): morphotectonic and seismic data. St Geol Camerti vol spec 1995(2):363–374
Ogniben L (1966) Hydrogeological features of Etna. Riv Min Sic 100–102:1–24
Parello F, D'Alessandro W, Aiuppa A, Federico C (2001) Geochemical mapping of Etna's aquifers. Publication No 2190 of the National Group for Defense against Hydrogeological Disasters. Officine Grafiche Riunite, Palermo
Ritmann A (1973) Structure and evolution of Mount Etna. Phil Trans R Soc London 274:5–16
Romano R, Sturiale C (1981) Geology of the southeastern slope of Etna, Sheet 270 IV (NO, NE, SO, SE). Boll Soc Geol It 100:15–40
Ruggieri R (2022) Karst of Sicily—a journey inside and outside the Island's Mountains. Springer Nature Switzerland AG 2022, pp XVIII, 474. https://doi.org/10.1007/978-3-031-07406-6
Schilirò F (1988) Methodological proposal for a geological-technical zoning of the built-up area of Maletto. Geol Tecnica 3(88):32–53
Stumm W, Morgan JJ (1996) Aquatic chemistry, chemical equilibria and rates in natural waters

3 Hydrogeological Basin of the Hyblean Mountains

Giovannella Pecoraino, Sabina Morici, Esterina Gagliano Candela, and Leonardo La Pica

3.1 Geological Framework

The basin of the Hyblean Mountains occupies the south-eastern extremity of Sicily and is a vast sub-circular plateau culminating at the centre at Mt Lauro, 987 m a.s.l., from which numerous offshoots radiate out in a gentle slope in all directions (Fig. 3.1).

The Hyblean Mountains are geologically made up of submarine volcanic expansions formed in the Neogene and raised together with thick limestone banks in the form of plateaus and shelves. Today, the Hyblean plateau is deeply incised by the gorges dug by the torrents, locally known as 'cave', long and deep gorges, squeezed between steep escarpments and limestone cliffs.

The alternation of limestone plateaus and quarries gives rise to a unique landscape, typical of the Hyblean Mountains, in which the limestone plateaus at the top, arid because of karst phenomena, alternate, in stark contrast, with the deep quarries. The latter are highly vegetated because of the constant presence of water from rivers and streams and the limited daytime sunshine.

The area of the Hyblean Mountains constitutes the northernmost sector of the African foreland, which, to the north and north-west forms the foredeep and beyond the Gela-Catania junction, is buried beneath the layers of the Gela Nappe.

G. Pecoraino (✉) · S. Morici · E. Gagliano Candela (Deceased) · L. La Pica
Istituto Nazionale di Geofisica e Vulcanologia, Sezione di Palermo, Via Ugo La Malfa 153, 90146 Palermo, Italy
e-mail: giovannella.pecoraino@ingv.it

S. Morici
e-mail: sabina.morici@ingv.it

L. La Pica
e-mail: leonardo.lapica@ingv.it

E. Gagliano Candela et al. (eds.), *Drinking Water in Sicily*,
https://doi.org/10.1007/978-3-031-60473-7_3

Fig. 3.1 The Hyblean Mountains. Imagine from Google Earth Pro v 7.3.6.9345 (Dicember 29, 2022). Sicily, Italy

Together with the submerged areas, this sector of the foreland is part of the Pelagian Block, which, as a whole, constitutes a stable zone extending from the Hyblean-Maltese escarpment to Tunisia, formed by a thick Meso-Cenozoic succession that is predominantly composed of carbonate with repeated interlayers of basic volcanites.

The Hyblean foreland, along its northern and western margins, is bordered by a foreland with siliciclastic sedimentation mainly fed by the northern quadrants during the Plio-Quaternary. This sector of the Plateau was affected by the Plio-Quaternary tectogenesis that produced the overlapping of the outermost front of the chain (Gela Fault) on the most peripheral parts of the foreland. This undercutting occurs with fault systems with a NE-SW trend on the northern edge, while the western margin is affected by a complicated system in which N-S or NNE-SSO directions (Scicli-Irminio river line) intertwine with NE-SE directions (Ispica line to the SE), and the Comiso-Chiaramonte system to the west. The latter is predominantly a carbonate plateau (Fig. 3.2) whose Triassic-Jurassic and partly Cretaceous levels are only known from subsurface data. The deepest strata reached in the subsurface is the Upper Triassic limestones and dolomites that are up to 4800 m thick.

The volcanic products present in the Hyblean area can be ascribed to three main events dated to the Upper Cretaceous, the Upper Miocene and the Plio-Pleistocene.

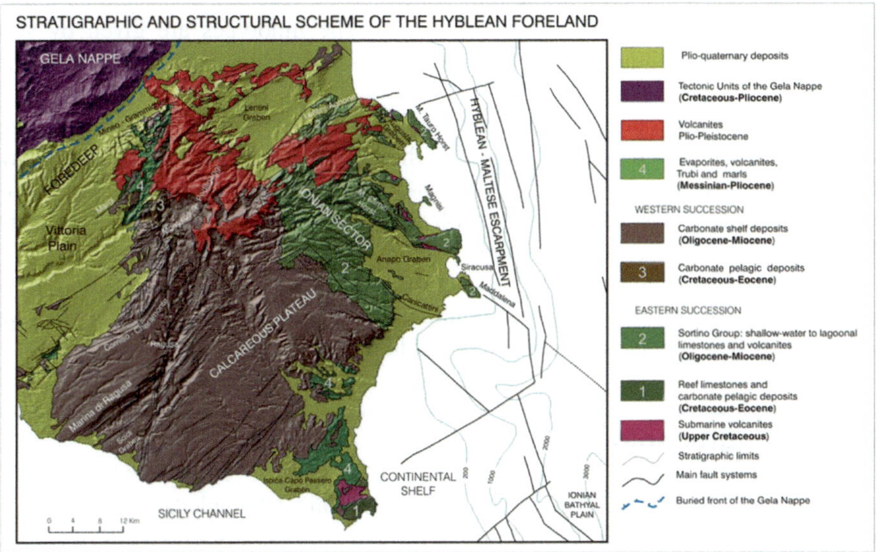

Fig. 3.2 Structural stratigraphic map of the Hyblean Mountains (modified after Ruggieri 2022)

On the surface, the oldest volcanic outcrops are of Cretaceous age, i.e. the ones near Pachino and those outcropping north of Siracusa.

Upper Miocene explosive volcanism produced breccias and volcanoclastic products that are distributed around diatreme-type craters. These outcrops occur in the northern Hyblean sector along a strip about 30–40 km wide extending in a NE-SW direction, while the most southerly outcrops occur at Monterosso Almo and along the Anapo basin as far as the Solarino area.

The Pliocene sediments are distributed discontinuously along the edges of the Hyblean plateau. Calcareous breccias, thin strips of "Trubi" and sands with Strombus Coronatus are present at the base of the lava covers of the Mt Lauro area, consisting of pillow-breccia below and subaerial lava flows above.

In the triangle between Vizzini, Licodia Eubea and Mineo, volcanic activity developed constantly in a submarine environment, as demonstrated by the presence of repeated levels of volcanic breccia alternating with "Trubi" and medium-Pliocene marls.

The Upper Pliocene is scarcely represented in these areas with the exception of the calcarenitic plateau of Licodia Eubea.

Submarine and subaerial volcanic activity seems to extend as far as the lower Quaternary, in the northernmost areas near the foredeep, where submarine volcanites are interspersed with levels of biocalcarenites from the lower Pleistocene.

The Quaternary deposits that rim the Hyblean Plateau belong to two main sedimentary cycles of infra- and mid-Pleistocene age. The Lower Pleistocene soils form a continuous belt around the limestone plateau, reaching considerable thicknesses along the northern and western edges. The lithotypes are given by soft yellowish biocalcarenites discordant on the Miocene substrate or on Pliocene volcanites.

The calcarenites change upwards and laterally to grey-blue clays reaching considerable thicknesses in correspondence with more depressed structures (Augusta). The second cycle is represented by the so-called Crotonian "bench" that discordantly covers terms of various ages with at its base a paelosoil originating from the period of the "Roman Regression".

On the western Hyblean border, along the extensive plain between Caltagirone and Vittoria, the infra-Pleistocene sedimentation is closed by a thick regressive sandy series that covers and sutures the Falda di Gela.

3.2 Hydrogeology

On the basis of geological-structural and geochemical knowledge, the Hyblean Mountains area can be subdivided into two main sectors: a South-Western sector, largely comprising the province of Ragusa, and a North-Eastern sector, largely coinciding with the province of Siracusa and to a lesser extent with the province of Catania (Fig. 3.3), which in turn are subdivided into sub-areas (water bodies) as listed below (Table 3.1):

South-western sector (two water bodies)

- Ragusano water body (carbonate)
- Vittoria Plain water body (sandy-sandy).

North-eastern sector (four water bodies)

- Lentinese water body (volcanic)
- North-east Siracusano (Syracusan) water body (mixed volcanic/carbonate)
- Southern Siracusano (Syracusan) water body (carbonate)
- Augusta-Priolo Plain (calcarenitic-sandy).

3.2.1 South-Western Sector "Ragusano"

This sector has been divided into two water bodies: the Ragusano water body and the Vittoria plain.

In the western part, consisting of the Comiso-Vittoria plain, there is a first aquifer in the Pleistocene calcarenitic-sandy terrains, at a medium depth (from 50 to 100 m) and a second deeper aquifer, in the carbonate substratum of the Ragusa Unit, confined by the marls of the Tellaro Unit. There is also a non-continuous

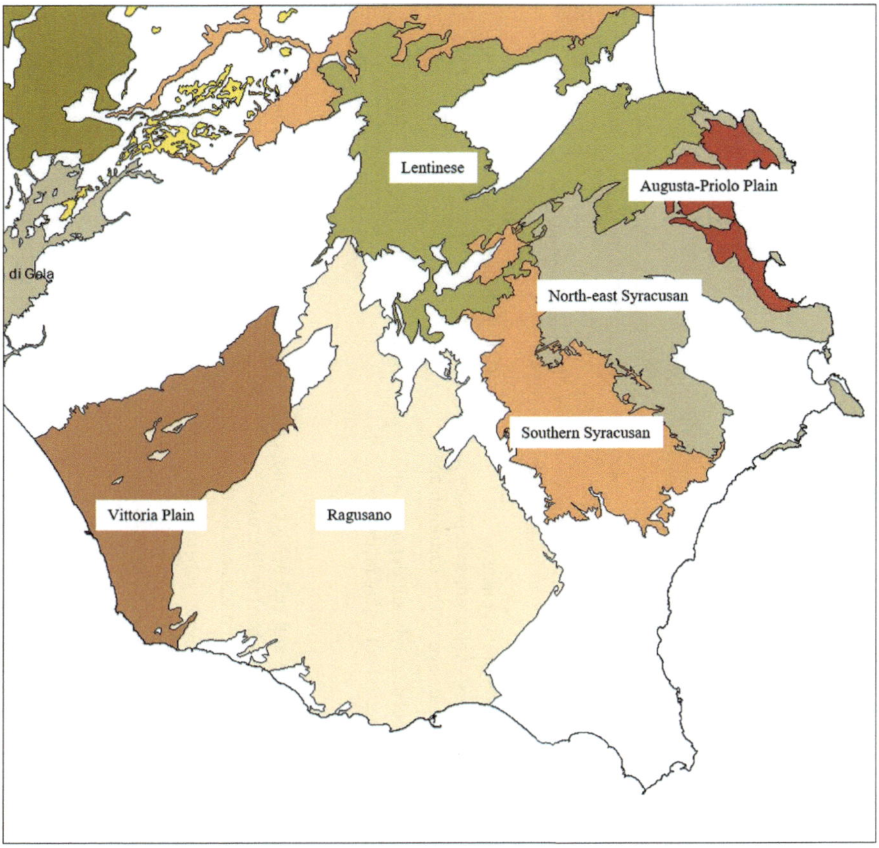

Fig. 3.3 Water bodies of the Hyblean Mountains basin

aquifer, interspersed between these aquifers, in the evaporitic series, characterised by water with a characteristic sulphurous odour and of little productive interest.

The supply of the western sector of the Vittoria structural depression comes both from effective infiltration and, especially at the border with the Hyblean plateau, from the carbonate massif. Some of the main Hyblean springs (Cifali, Passolatello, Diana, etc.) are found in this transitional zone, emerging from the over-imposed threshold and having the limestone plateau as their feeding area.

In the south-western sector, mainly affected by the carbonate outcrops of the Ragusa Unit, there is a first aquifer, partially confined, in the calcarenitic series of the Irminio Member, at an average depth between 100 and 150 m, followed, separated by a variable thickness of marly-clayey terms, by a deeper and more productive confined aquifer in the calcareous-marly series of the Leonardo Member of the aforementioned Unit. In both cases the permeability is by fracturing, as the contribution of the porosity of the limestone itself is almost nil. Locally, the presence of important regional tectonic structures puts the two aquifers in hydraulic

Table 3.1 Water bodies and their description

Water body	Hydrogeological complex	Mouton hydrogeological complex	Description	PTA hydrogeological complex	Type of aquifer
North-eastern Siracusano	CA Sandstone	Calcareus complex	Mixed volcanic/carbonate aquifer	Calcareous complex and syracuse volcanite complex	Multilayer aquifer
Ragusano	CA Sandstone	Calcareus complex	Carbonate sequence	Calcareus complex	Multilayer aquifer
Southern Siracusano	CA Sandstone	Calcareus complex	Succession of carbonate, calcareous—calcarenitic deposits from the Cretaceuos age to the Tortonian age	Calcareus complex	Multilayer aquifer
Vittoria plain	DET/AV plio-quaternary detritic formation valley alluvials	Complex of sand/ complex of alluvial deposits of major courses water	Pleistocene sandy-silt deposits with interspersed silty-clay layers	Sandy—calcarenitic	Single layer
Augusta -Priolo plain	DET/DQ plio-quaternary detritic formation Alluvial of quaternary depressions	Complex of sands	Coarse sands and yellowish organogenic calcarenites (bench) with cross stratification often terraced on the top		Single layer
Lentinese	VU volcanites	Complex of pyroclastites, lavas and volcanites in general	Product of volcanic, subereal and underwater activities occurring in the Pliocene and lower Pleistocene	Volcanite complex	Single layer

contact, while in some cases the effect of a significant hydraulic load determines the artesian emergence of the deep aquifer.

Along the coastal strip and up to the eastern border with the territory of Syracuse, the Ragusa limestones present an aquifer with fair to good potential, confined by the marls of the Tellaro Unit.

Finally, aquifers in alluvial deposits, conoids and recent sands are of low to medium productivity because they are generally poorly developed, both horizontally and vertically. The vulnerability of the aquifers in the carbonate series is generally high, especially as they are not protected above by clayey-marly soils and, more generally, because they are characterised by high permeability due to karstification.

The karst dissolution phenomena within the network of fractures in the Ragusa limestone began from the earliest phases of uplift of the plateau, from the end of the Tertiary period, and continued in the emerged areas throughout the Quaternary period, generating an initial epikarstic drainage system followed by increasingly articulated systems of conduits and tunnels. The deepening of the hydrokarstic circuits, conditioned by both tectonics and Pleistocene climatic variations (eustatic regressive and transgressive phases) is evidenced by a series of fossil conduits at various levels in the Ragusa limestone series.

3.2.2 North-Western Sector

The north-eastern sector of the Hyblean Mountains for hydrogeological purposes can in turn be subdivided into four water bodies: the Lentinese basin, the north-eastern Siracusano, the southern Siracusano and the Augusta-Priolo plain (Fig. 3.2). The water bodies present different geochemical characteristics in relation to the directions of groundwater flow. In particular, in the northern portion, from Mt Lauro to the Lentini Plain, groundwater circulates predominantly in the Plio-Pleistocene volcanic deposits with a north-northeast direction of flow. The semi-permeable substrate of this aquifer consists locally of Miocene volcanites that are often altered in the upper part by argillification processes. A structural high along the NE-SW alignment separates this body of water from the adjacent mixed aquifer (Augusta basin), in which the alternation of deposits of volcanic origin with soils of the carbonate succession is more marked. Further to the west is the "Siracusano" carbonate basin, bordered to the north by the Melilli-Climiti Mountains graben, a structural high with a WNW-ESE direction. In this basin, water flows mainly towards the SW. The main aquifer involves the limestone of the Palazzolo Unit and the Mt Climiti Unit. This carbonate series rests to the west on the Miocene marls of the Tellaro Unit; in the area of Siracusa-Solarino and Cassibile the limestones are covered by Plio-Pleistocene sediments.

The carbonate aquifer has transmissivity values between 0.1 and 9×10^{-3} m^2/s and good permeability, in relation to the widespread karstification processes that have affected the Miocene carbonate series since its emergence.

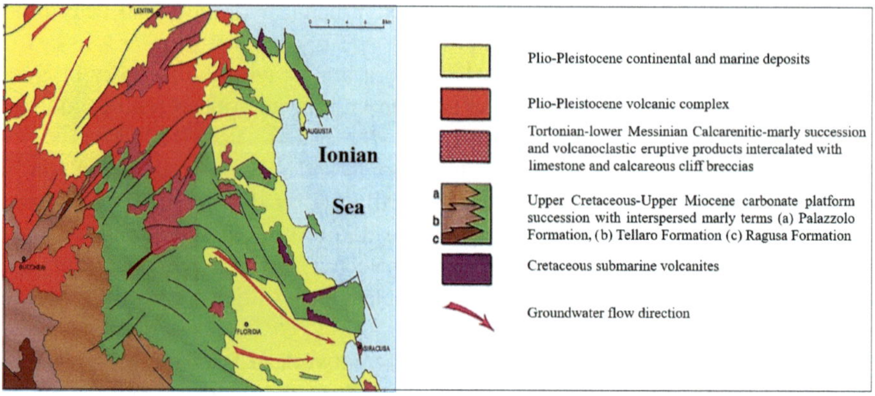

Fig. 3.4 Hydrostructural map of the north-eastern Hyblean sector (modified from Lentini et al. 1984)

The water body of the Augusta-Priolo plain consists of coarse sands and yellowish organogenic calcarenites (bench) with cross stratification (Fig. 3.4) often terraced at the top. The substratum is made up of clays with a thickness varying from a few to over 270 m. In some places the clays of the substratum are missing and the sands and calcarenites rest directly on the lower permeable strata, for which, due to their high permeability, they do not constitute any protection but with which they are in hydraulic continuity. The maximum thickness of the sands and calcarenites is slightly more than 20 m. They are present all along the gulf of Augusta and inland up to a maximum altitude of 200 m. They are the closing sediment of the gulf: the deposits that filled the large graben existing in the lower Miocene units.

The water table contained in these sediments is fed exclusively by local infiltration and largely drained by the underlying watercourses.

3.3 Hydrogeochemical Characterisation

In this hydrological basin 111 springs are exploited for potable use (PRGA database, INGV 2005), from which 13 have been selected (Fig. 3.5) and sampled for this study and listed in Table 3.2. Selected springs fall in three water bodies: Lentinese (Favara and Paradiso), Ragusano (Passolatello, Cifali, Scalarancio, Misericordia 2, Oro, Corchigliato, Scianna Caporali, Fontana Grande, Cilone, Cava Volpe) and Southern Siracusano (Gelso Steartino, San Giovanni Lo Vecchio) (Table 3.2 and Fig. 3.5).

The sampled waters show limited ranges in variation in almost all measured physic-chemical parameters. The temperature ranges between 14 and 19 °C, the pH indicates the presence of neutral waters with values between 7.12 and 7.81. Finally, the electrical conductivity (at 20 °C) ranges from 396 to 597 µS/cm.

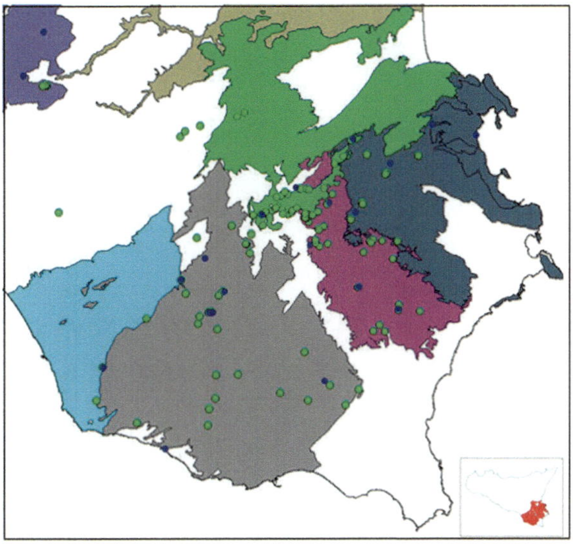

Fig. 3.5 Map of the Hyblean Mountains hydrological basin. The location of springs is shown (green dots represent PRGA spring, bleu dots spring sampled for this work). Colours identify water bodies: light blue—Vittoria Plain; light gray—Ragusano; Purple—Southern Siracusano; dark gray—north-east Siracusano; green—Lentinese)

In the Langelier-Ludwig diagram (Fig. 3.6), all the sampled spring waters fall within the quadrant of bicarbonate-alkaline-earthy waters. In the diagram, literature data are shown for comparison.

Water circulating in the carbonate and volcanic water bodies has a prevalent bicarbonate-earth alkaline connotation (Fig. 3.6), due to dissolution processes of calcium carbonates (Calcite) and calcium and magnesium carbonates (Dolomite). It is possible to identify a trend evolving respectively towards a chlorinated-sulphate-alkaline component probably linked to ion exchange phenomena with the rocks.

The ternary diagrams (Fig. 3.7) confirm the limited compositional variability observed in Fig. 3.6, with a more or less marked tendency to the vertices of sodium and potassium. Again, it can be seen that the influence of seawater alone cannot be attributable to this evolution, thus strengthening the hypothesis of chemical processes of ionic exchange. The slight differentiation of the bivalent ions (Ca and Mg) with a tendency of some samples towards an evolution to waters richer in magnesium could indicate an interaction with rocks of a more exquisitely dolomitic or, in any case, more magnesian type.

3.4 Isotopic Characteristics of the Catchment Area

The drinking water sampled for this study in the Hyblean basin aligns on a line very close to the GMWL and MMWL (Fig. 2.6). The LMWL* is $\delta^2H = 6.5\ \delta^{18}O + 9.87$ (Grassa et al. 2006). The stable isotope composition of groundwater ranges from –6.47 to –5.87 δ‰ for $\delta^{18}O$ and from –34 to –28 δ‰ for δ^2H (INGV 2005; Grassa et al. 2006).

Table 3.2 Chemical and isotopic composition of the sampled waters

	T	EC	pH	Eh	Na	K	Ca	Mg	Cl	NO_3	SO_4	HCO_3	dD	$d^{18}O$	WB
Paradiso	16.3	416	7.53	76	0.57	0.043	3.65	0.82	0.45	0.23	0.22	4.39	−32	−6.07	LE
Favara	13.9	597	7.46	165.8	0.97	0.038	4.19	1.48	0.97	0.27	0.73	4.7	−34	−6.01	LE
Oro	16.4	475	7.12	219	0.65	0.055	4.98	0.38	0.771		0.21	4.7	−34	−6.45	RG
Misericordia	16.5	450	7.19	211	0.47	0.027	4.73	0.32	0.52	0.61	0.19	4.56	−34	−6.45	RG
Corchigliato	15.9	436	7.37	223	0.43	0.02	4.49	0.47	0.5	0.27	0.27	4.46	−34	−6.47	RG
Cifali	17.0	451	7.34	162	0.48	0.03	3.85	0.9	0.57	0.56	0.25	4.1	−36	−6.43	RG
Passolatello	19.4	527	7.58	172	0.86	0.047	5.03	0.43	0.92	1.41	0.22	4.61	−32	−6.07	RG
Scalarangio	19.0	565	7.15	135	0.67	0.042	6.00	0.36	0.83	1.71	0.23	5.0	−28	−5.33	RG
Gelso Steartino	16.1	515	7.22	203	0.48	0.024	5.21	0.31	0.52	0.61	0.27	4.75	−30	−6.08	RG
S. Giovanni Lo Vecchio	17.0	492	7.26	194	0.46	0.012	4.43	0.3	0.52	0.59	0.18	4.1	−30	−5.87	RG
Scianna Caporali	17.1	396	7.4	291	0.52	0.023	3.73	0.54	0.62	0.7	0.18	3.7	−34	−6.28	RG
Cava Volpe	17.2	430	7.81	243	0.41	0.023	4.62	0.43	0.47	0.34	0.23	4.6	−34	−6.19	RG
Fontana Grande	16.4	577	7.29	254	0.7	0.05	5.55	0.32	0.71	1.22	0.37	5.0	−31	−6.27	SR
Cilone	16.5	538	7.25	120	0.55	0.041	5.3	0.31	0.65	1.1	0.16	4.8	−31	−6.15	SR

Analytical data are expressed as meq/L, EC in μS/cm and δ^2H and $\delta^{18}O$ in ‰ versus SMOW. WB is for water body. LE is Lentinese, RG is Ragusano and SR is southern Siracusano

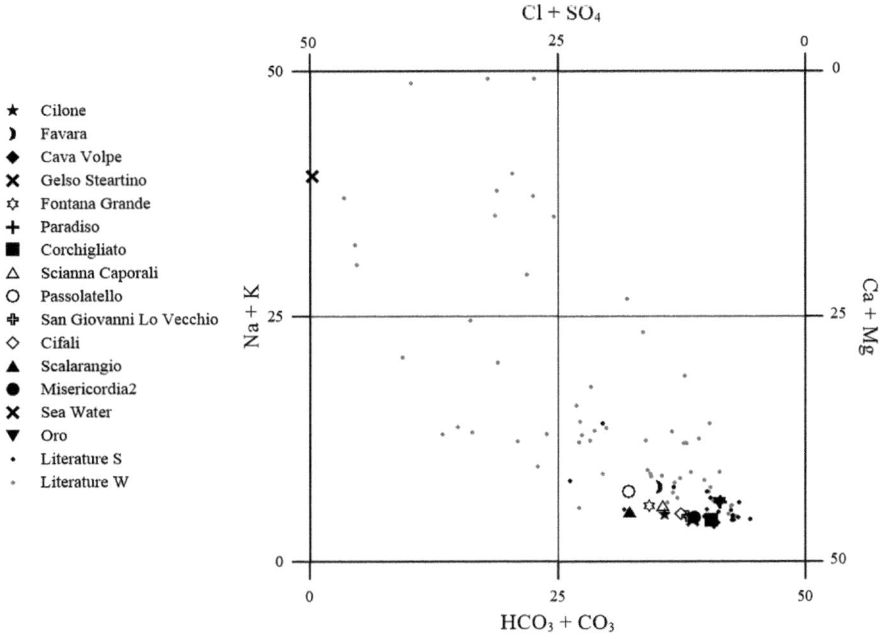

Fig. 3.6 Langelier–Ludwig classification diagram. The spring compositions fall in the bicarbonate-alkaline-earthy quadrant. Literature data are shown for comparison (INGV 2005). S: spring, W: well

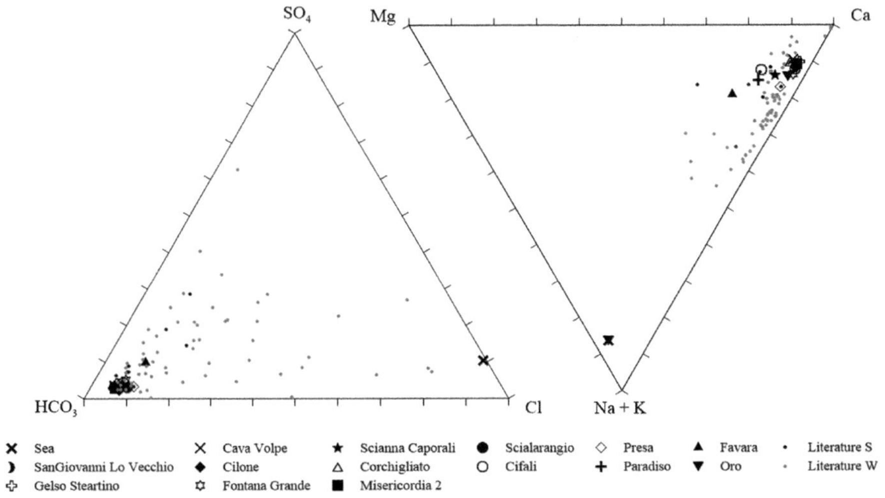

Fig. 3.7 Anions (left) and cations (right) ternary diagrams for the Hyblean Mountains catchment area. S: spring, W: well

The weighted average meteoric water line and the general trend of the points of all water bodies that lie on a virtual line between this line and the seawater composition show that the various water bodies are fed by local meteoric water with slight inputs of water from higher altitudes (i.e. by more negative values) and more or less abundant mixing with seawater (Fig. 3.7). Grassa et al. (2006) hypothesised, on the basis of $\delta\ ^{2}H$ and $\delta^{18}O$ isotope data and Cl depositional rate in rainwater, an origin of rainwater moisture from Mediterranean Sea.

Groundwater values plot below the LMWLs (Fig. 3.8) confirming that evaporation processes occurred during rainfall or post infiltrating processes as mixing with evaporated waters (Grassa et al. 2006; Liotta et al. 2013).

The evaporative processes, inducing a positive $\delta^{18}O$ shift, are highlighted in Fig. 3.9 where points plot above the line representative of the $\delta^{18}O$—Altitude relationship (isotopic gradient of 0.16 δ‰/100m, Table 1.7 relative to sector 2).

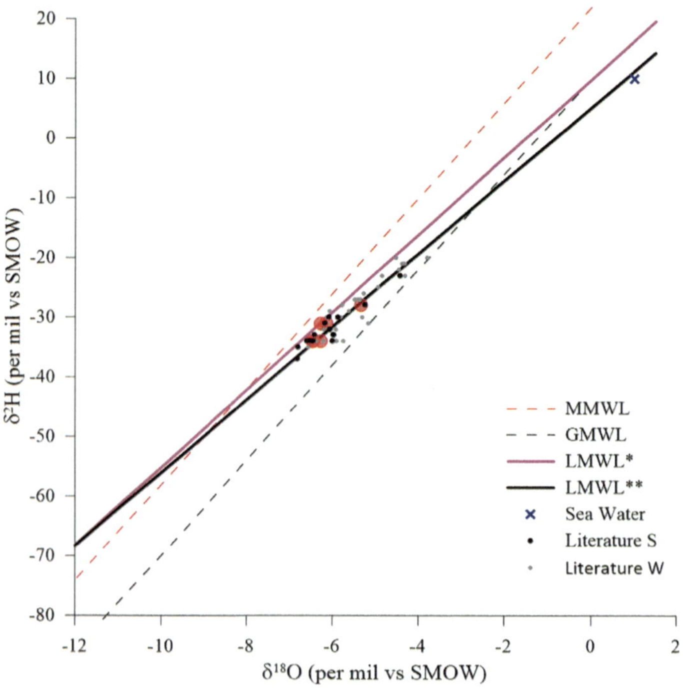

Fig. 3.8 $\delta^{2}H$—$\delta^{18}O$ (in ‰ with respect to SMOW) diagram of spring waters. The LMWL* is from Grassa et al. (2006), LMWL** is from Liotta et al. (2013). The GMWL, MMWL and literature data are shown for comparison. S: spring, W: well

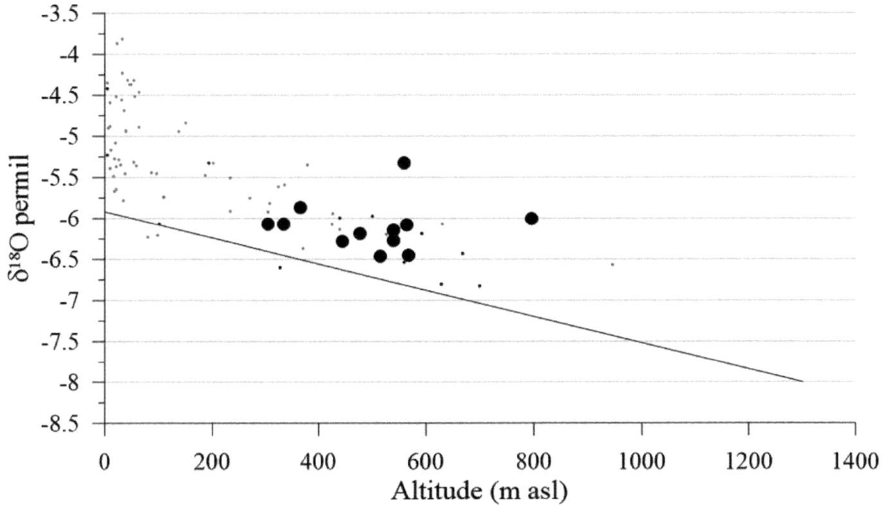

Fig. 3.9 Relationship between $\delta^{18}O$ and altitude

3.5 Trace Elements

Trace elements in the Hyblean basin, with some exception, do not show wide ranges in variation per element (Table 3.3 and Fig. 3.10) despite flowing in different and geologically diverse water bodies (limestones and volcanites).

Some elements are preferentially enriched in the carbonate aquifers (Ba, U), whereas others are more abundant in volcanic aquifers (Fe, Zn, V). All elements have concentrations below the legal limits for drinking water.

Table 3.3 Trace element composition of sampled waters

	Li	B	Al	V	Cr	Mn	Fe	Co	Ni	Cu	Zn	As	Se	Rb	Sr	Sb	Ba	Pb	U	WB
Paradiso	nd	17.2	2.14	6.59	0.62	0.78	14.2	nd	1.95	2.07	4.58	0.85	0.49	nd	510	0.13	29.4	0.216	1.320	LE
Favara	nd	14.1	0.88	13.2	0.92	0.1	19.8	nd	1.98	0.8	0.44	1.21	0.88	nd	519	bdl	17.9	0.159	1.640	LE
Oro	0.84	20.3	bdl	1.47	0.31	bdl	2.5	0.11	1.95	0.9	0.87	4.01	0.56	1.00	431	bdl	25.5	bdl	1.260	RG
Misericordia	0.92	16.1	bdl	1.48	0.34	bdl	2.2	bdl	1.9	0.48	0.57	3.57	0.7	0.581	401	bdl	22.8	bdl	1.360	RG
Corchigliato	1.23	13.9	bdl	1.87	0.17	bdl	2.1	bdl	2.09	0.56	0.56	3.56	0.91	0.602	529	0.07	21.6	bdl	1.520	RG
Cifali	nd	16.4	0.58	2.77	0.48	bdl	13.7	nd	2.14	0.79	0.98	1.11	1.41	nd	766	0.1	40.1	0.086	1.280	RG
Passolatello	0.9	26.8	0.32	1.5	0.33	0.11	2.7	0.12	1.8	1.27	1.63	3.53	0.55	0.532	362	bdl	38.4	0.139	0.821	RG
Scalarangio	0.82	22.8	0.22	2.14	0.15	0.53	1.8	0.13	1.95	0.49	1.11	0.35	0.46	0.523	340	bdl	28.1	bdl	1.100	RG
Gelso	nd	16.8	1.03	0.93	0.29	bdl	24.2	nd	2.15	0.45	0.34	0.62	0.41	nd	463	bdl	12.6	0.091	1.220	SS
S. Giovanni	0.34	15.9	1.16	1.12	0.25	bdl	2.3	0.11	1.6	0.42	0.74	0.19	0.28	0.218	>500	bdl	8.5	bdl	0.859	SS
Scianna Caporali	1.25	19.4	8.04	2.56	0.74	bdl	5.8	0.28	0.32	0.19	1.00	0.67	1.1	0.495	641.7	0.05	31.4	0.01	0.885	RG
Fontana Grande	0.51	31.8	12.53	2.56	0.57	0.236	6.5	0.51	1.03	0.59	1.45	0.5	0.33	0.864	351.3	0.05	17.7	0.013	1.270	RG
Cilone	0.68	18.7	2.19	2.09	0.55	0.14	4.9	0.27	0.21	bdl	0.34	0.49	0.28	0.824	511.4	0.04	21.5	bdl	1.251	RG
Cava Volpe	1.21	15.7	4.22	1.37	0.28	bdl	5.4	0.35	0.85	0.39	1.59	0.42	1.08	0.719	423.2	0.08	22.7	0.012	1.430	RG

Concentrations are in μg/L. WB is for water body. LE is Lentinese, RG is Ragusano and SS is southern Siracusano (D'Agostino et al. 2021)

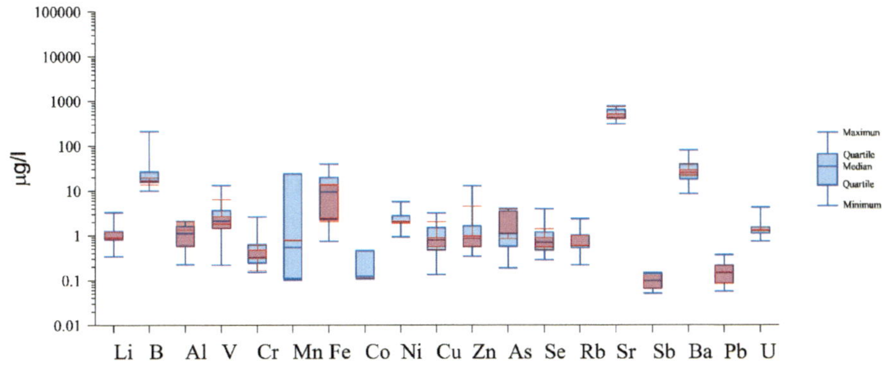

Fig. 3.10 Box-whisker plot of trace element at Hyblean Mts Basin. Red box spring data (this work). Light blue box data from INGV (2005)

Bibliography and Recommended Further Reading

Beccaluva L, Di Grande A, Lo Giudice A, Macciotta G, Siena F (1993) Geopetrographic map of the central-northern Hyblean area, south-eastern Sicily. Scale 1:50 000. Università di Catania, S. EL. CA., Firenze

D'Agostino F, Avellone G, Ceraulo L, Di Stefano V, Indelicato S, La Pica L, Morici S, Vizzini S, Bongiorno D (2021) Groundwater of Sicily (Italy) close to landfill sites: quality and human health risk assessment. Exposure Heal. https://doi.org/10.1007/s12403-021-00401-z

Dansgaard W (1964) Stable isotopes in precipitation. Tellus 16:436–468

Ferrara V (1970) Hydrogeology of the Francofonte area (Syracuse). Atti Conv Intern Sulle Acque Sotterr, 6–8 Dic 1970:287–311

Gat JR, Carmi H (1970) Evolution of the isotopic composition of atmospheric waters in the Mediterranean Sea area. Geophys Res 75:3039–3040

Grassa F, Favara R, Valenza M (2006) Moisture source in the Hyblean Mountains region (south-eastern Sicily, Italy): evidence from stable isotopes signature. Appl Geochem 21:2082–2095

Grasso M (1997) Geological map of the central-southern sector of the Hyblean plateau (Pr. Ragusa, south-eastern Sicily). Scale 1:50,000. S.EL.CA., Firenze

Istituto Nazionale di Geofisica e Vulcanologia (INGV), Sezione di Palermo (2005) First characterisation and monitoring of underground waters for the drafting of the 'Water Protection Plan of the Region of Sicily'. Per conto della Regione Siciliana-Commissario Delegato per l'Emergenza Rifiuti e Tutela delle Acque

Lentini F (1986) Geological map of the north-eastern Hyblean sector, scale 1:50 000. Università di Catania, Istituto di Scienze della Terra, S.EL.CA., Firenze

Lentini F, Bommarito S, Carbone S, Cugno G, Di Geronimo I, Grasso M, Iozzia S, La Rosa N, Romeo M, Scamarda G, Sciuto F (1984) Geological map of south-eastern Sicily. Scale 1:100 000. Università di Catania, Istituto di Scienze della Terra. S.EL.CA., Firenze

Lentini F, Carbone S, Catalano S, Grasso M (1996) Elementi per la ricostruzione del quadro strutturale della Sicilia orientale. Mem Soc Geol It 51:179–195

Liotta M, Grassa F, D'Alessandro W, Favara R, Gagliano Candela E, Pisciotta A, Scaletta C (2013) Isotopic composition of precipitation and groundwater in Sicily, Italy. Appl Geoch 34:199–206

Longaretti G, Rocchi S (1992) Il magmatismo dell'Avampaese Ibleo (Sicilia orientale) tra il Trias e il Quaternario: dati stratigrafici e petrologici di sottosuolo. Mem Soc Geol It 45:911–925

Regione Siciliana (1987) Regional Water Reclamation Plan, Land and Environment Department
Ruggieri R (1997) The Karst in the Hyblean Mountains (south-eastern Sicily). Atti del Convegno "Il carsismo nell'area mediterranea". Supplemento al n. 23 di Thalassia Salentina, 47–55
Ruggieri R (2022) Karst of Sicily—a journey inside and outside the island's mountains. Springer Nature Switzerland AG 2022, pp XVIII, 474. https://doi.org/10.1007/978-3-031-07406-6
Ruggeri R, Grasso M (2000) Stratigraphic and structural characteristics of the Hyblean Ragusa plateau and its implications on karst morphogenesis. Atti 1° Seminario di Studi sul carsismo negli Iblei e nell'area sud Mediterranea. Speleologia Iblea 8:19–35

Madonie Hydrogeological Basin

4

Esterina Gagliano Candela, Sabina Morici, Leonardo La Pica, and Giovannella Pecoraino

4.1 Geological Framework

The Madonie Mountains are the most important mountainous complex in western Sicily. They are conventionally limited to the east by the Fiumara di Pollina and to the west by the northern sector of Imera river.

The Madonie area, located in the eastern sector of the province of Palermo (Fig. 4.1), is characterised by the presence of predominantly calcareous-dolomitic reliefs.

The Pizzo Carbonara group, with its peak at 1979 m a.s.l. second in altitude after Etna, constitutes the main orographic structure of the Madonie complex. The central nucleus of the carbonate massif crops out at the summit, in correspondence of Pizzo Carbonara, Mt Ferro and Pizzo Antenna, a karstic plateau at altitudes above 1600 m a.s.l.. The major reliefs are surrounded by vast clay and sandstone plains that create rounded hills and mountains or elongated ridges. The hydrogeological basin of the Madonie falls within the catchment areas of the northern Imera, Roccella, Pollina and southern Imera rivers.

The Madonie karst landscape is characterised by both epigean karst forms (dolines, polje, blind valleys, sinkholes and furrowed fields or karren) and hypogean forms (sink holes, karst wells, caves). The presence of sinkholes in

E. Gagliano Candela (Deceased) · S. Morici · L. La Pica · G. Pecoraino (✉)
Istituto Nazionale di Geofisica e Vulcanologia, Sezione di Palermo, Via Ugo La Malfa 153, 90146 Palermo, Italy
e-mail: giovannella.pecoraino@ingv.it

S. Morici
e-mail: sabina.morici@ingv.it

L. La Pica
e-mail: leonardo.lapica@ingv.it

E. Gagliano Candela et al. (eds.), *Drinking Water in Sicily*,
https://doi.org/10.1007/978-3-031-60473-7_4

Fig. 4.1 The Madonie Mountains. Imagine from Google Earth Pro v 7.3.6.9345 (December 29, 2022). Sicily, Italy

numerous dolines in the Madonie suggests a connection between epigean and hypogean forms that conditions the underground water circulation. The intense karstification of the carbonate terrain influences surface and underground water circulation, since the presence of underground karst circuits set along preferential dissolution lines, favours the rapid flow of water along tectonic lines.

The Madonie Mountains are located in the central sector of the northern Sicilian chain and are the result of the Miocene tectonic superimposition of various carbonate, carbonate-silicoclastic and terrigenous geological bodies of Upper Triassic-Lower Miocene age, generally with a southern orientation. The tectonic strata, which constitute the Madonie sector of the Sicilian chain, reached their current reciprocal relationships between the beginning of the Miocene and the beginning of the Pliocene. These bodies derive from the deformation of a pelagic basin and escarpment succession (Autochthonous Imerese Dominio, thereafter Autochthonous is Auct.), and of a carbonate platform succession (Auct. Panormide Dominio) identified during the Mesozoic extension phases and overlain by silicoclastic terrigenous deposits (Auct. Numidian Flysch). Subordinate are the outcrops, reduced to isolated slabs, of pelitic deposits (Varicoloured clays Auct.) and calcareous-marly deposits (Polizzi and Tusa Unit) of the Auct. Sicilide Domain. The major reliefs are generally surrounded by terrigenous and fluvio-deltaic soils (Auct. Cozzo Terravecchia Unit) and, subordinately, by deposits of the Messinian evaporitic cycle and Pliocene ‘Trubi’.

In the Middle-Upper Pliocene, further deformation of the tectonic edifice is associated with transpressive fault systems, oriented NE-SW, which determined the extrusion of geometrically lower tectonic bodies into the structural edifice, producing, in addition, retrovergent out-of-sequence systems. The structural analysis of the entire tectonic edifice of the Madonie has made it possible to recognise a stack of superimposed tectonic units, arranged as follows, from bottom to top:

- Imeresi Unit (*Mt dei Cervi Unit, Mt Quacella Unit*);
- Panormid Units (*Pizzo Dipilo-Pizzo Carbonara Unit*);
- Sicilidi Units, which tectonically overlie the Numidian Flysch that constitutes the terrigenous cover of both the previously mentioned units;
- Post-orogenic deposits of the Upper Tortonian—Lower Pliocene (terrigenous, evaporitic and carbonate), which discordantly cover the underlying stratum structure.

4.2 Hydrogeology

The Madonie area is the most important water reservoir in central-northern Sicily due to its high annual rainfall and particular hydrogeological structures. The various similar lithological types outcropping in the Madonie sector have been grouped into homogeneous complexes from a hydrogeological point of view:

Calcareous-marly and clay-arenaceous-calcareous complex (Mufara Unit)
It is essentially made up of clays and variegated marls with interspersed calcilutitic, calcisiltitic and calcarenitic levels and dolomitic breccias. This complex, as a whole, has a very low permeability (ranging from 10^{-7} to 10^{-9} m/s), which is further reduced when it is intensely tectonised. Vulnerability is low and practically negligible.

Calcareous-dolomitic complex
Various lithological terms with a similar permeability have been grouped together in this complex. These are the calcilutites and calcarenites of the Auct. Scillato Unit (thereafter Un.) of the Upper Trias, the dolomitic breccias, dolorudites and doloarenites of the Auct. Fanusi Un., of the Lower Lias and the large resedimented carbonate bodies (essentially calcarenites and calcirudites) of the Auct. Crisanti Un., as well as the dolomites and dolomitic breccias of the Quacella Un. The considerable exposure of the outcrops allows a large amount of water to infiltrate, generating high flow sources. Vulnerability is high, depending on the higher or lower degree of karstification of the limestone. The prevailing permeability is high due to fracturing and, in places, increased porosity.

Predominantly limestone complex
In this complex, the terms limestone and limestone-dolomite have been unified, pertinent to the successions deriving from the deformation of the Panormide Platform domain. Its hydrogeological characteristics make it the main aquifer of the Madonie territory. This complex appears discretely stratified and intensely fractured and karstified, and is characterised by a high degree of relative permeability due to cracking and epigean and hypogean karstification (Agnesi et al. 1998), allowing high rates of effective infiltration. Even in this complex, fracturing develops in several systems; the main one generally presents a trend almost perpendicular to the secondary one. The vulnerability of the complex is considered to be very high, but the modest anthropization of the higher altitude areas and the lack of widespread or point pollution source risk centers greatly limit it.

Marly limestone and silico-marly complex
Grouped in this complex are the sequences of siliceous argillites, radiolarites and radiolarian marls of the Auct. Crisanti Un., the calcilutites and calcisilites with intercalated banks of graded biocalcarenites (Auct. Caltavuturo Un.), the calcilutites, calcisilites and marls with flint strips ("Scaglia" Auct.) and the calcilutites, calcarenites and marls of the Auct. Polizzi Un.

Cracking permeability is low and this increases locally in calcarenite successions. Vulnerability is considered medium to low.

Meso-Cenozoic carbonate megabreccia complex
This complex is mainly composed of calcareous megabreccias and olistolites arranged in lenticular or stratiform bodies with a thickness that varies from a few centimetres to several tens of metres and interspersed with various levels in the clayey successions of the Numidian Flysch, such as in the Fosso Canna gallery. Porosity is almost nil, while cracking is linked to their degree of tectonisation. Permeability varies from 10^{-7} to 10^{-8} m/s. The modest outcrops generate small springs of ephemeral flow. Vulnerability is medium, but extremely localised.

Numidian Flysch arenaceous clay complex
The complex includes the pelitic-arenaceous sequences of the Numidian Flysch: sandstones with intercalations of clays and conglomeratic sandstones. Permeability is created essentially by various degrees of cracking at the sandstone levels. The prevalence of arenaceous terms over pelitic terms results in a groundwater circulation, which, although discontinuous, is, on the whole, of some interest. Vulnerability is low and practically negligible.

Predominantly sandstone complex
The conglomeratic sandstone sequences of the Numidian Flysch and the arenaceous sequences of the Reitano Flysch are grouped in this complex. Permeability is modest and related to porosity, locally higher due to the presence of diffuse cracking. Due to the presence of anthropic settlements (like as livestock farms), the vulnerability, generally lower for these aquifers, is to be considered medium.

This complex hosts small discontinuous aquifers in a fracture network. The most significant springs are Camparelle (~8 l/s), Centomasi (~8 l/s), and Piano Lana.

Predominantly clayey-marly complex

In this complex the numerous units pertaining to the Sicilid and Oligo-Miocene Units are united, due to obvious lithological and hydrogeological affinities. In fact, the following are represented in this complex: the Serravallian-Tortonian clays (Pre-Terravecchia Auct.), the marls and sandy clays of the Tortonian (Auct. Terravecchia Un.), the thinly stratified clays, siltstones and sandstones of the Numidian Flysch (pelitic facies), the Auct. Varicolored Clays, frequently tectonised, sometimes chaotic, and incorporating into a more or less abundant sandstone fraction, the alternating levels of arenaceous and clayey-marly levels of the Tusa volcanoclastites (Tusa Un. or 'Tufiti di Tusa' Auct.), and the clayey-marly and calcarenitic and calcisylitic levels of the Un. Gratteri Auct. These various composite unit act as impermeable relative with respect to all the other hydrogeological complexes in the Madonie sector. Vulnerability is practically zero, except in localised situations, due to the presence of small springs or wells.

Conglomerate complex

Polygenic conglomerates, mostly pebbly, more or less cemented, with mainly carbonate and/or arenaceous clasts and a sandy matrix, are found in the southern and western sectors of the Madonie. Permeability is generally medium to low and ranges from 10^{-5} to 10^{-6} m/s. Vulnerability is considered medium to low with local conditions of higher sensitivity to point source pollution.

Calcareous chalky-marly complex

This complex includes all the units pertaining to the evaporitic series and the marly and marly-calcareous units of the Auct. 'Trubi' of the Lower Pliocene. In this complex are grouped: marls and calcareous marls ("Trubi"), gypsum arenites alternating with banks of macrocrystalline gypsum with clay intercalations, calcarenites and organogenic limestones biolithic to Porites (Baucina Auct. Un.).

The waters contained in this complex are in part highly mineralised, and generally more or less poor in quality and therefore not usable for drinking water purposes. The vulnerability of the evaporitic complex is medium to low: it is higher where chalks and/or limestones are present in the outcrop and lower or almost negligible where the "Trubi" mask the underlying evaporitic terms.

Alluvial complex and littoral deposits

The alluvial complex, outcropping in the valley floor and along the river mouths, consists of lenticular bodies made up of polygenic materials, from coarse to fine. The different degrees of permeability makes it possible to define this complex as a multi-bed complex, since the latter, housed in these materials, is irregularly subdivided into a series of overlapping bodies of water whose piezometric height is quite variable.

The littoral deposits extend widely along the stretches of the coastline, most frequently near the mouths of the Imera and Pollina rivers. In these stretches, there are wide bands of sandy soils of fluvial origin that have been reworked by the sea, sometimes arranged in a succession of rope-shaped structure parallel to the coast. Fine sands predominate, with some presence of very fine fractions or even episodes of marshy origin that greatly lower the permeability by porosity, thus hindering the entry of seawater inland.

The vulnerability of alluvial aquifers is high, not only because of their high average permeability, but also because of the conditions of anthropic exploitation to which they are subjected, especially in the valley floors and along the coast.

Aquifer detritus and eluvial-colluvial covers

Along the flanks of the main carbonate reliefs and at the foot of the thick quartzarenitic banks of the Numidian Flysch detrital materials and sometimes eluvial products and colluvial deposits can be found. The detrital materials are essentially made up of heterometric clasts, even of large dimensions. Permeability is generally medium–low, from 10^{-5} to 10^{-7} m/s and drainage is very rapid.

The detrital deposits have an average permeability, which is higher than that of alluvial deposits, but their position and morphological layout, the rarity of the road network, the almost total lack of residential or only occasional settlements, the absence of particular crops, except for some modest grazing, make the risk of pollution of these units rather modest.

4.2.1 Hydrogeological Outline

The relationships between the various hydrogeological structures identified in the Madonie have been summarised in the hydrogeological scheme of Fig. 4.2 (Cusimano 1987; Aureli et al. 2001). Within the Madonie hydrogeological basin, it is possible to distinguish four autonomous hydrogeological structures:

- *Mt dei Cervi hydrogeological unit;*
- *Mt Quacella hydrogeological unit;*
- *Pizzo Carbonara-Pizzo Dipilo hydrogeological unit;*
- *Pizzo Catarineci hydrogeological unit.*

The first two hydrogeological units consist of a lithostratigraphic succession belonging to the Imerese domain, the third is formed by lithostratigraphic units of the Panormide domain, and the fourth by conglomeratic arenaceous sequences of the Numidian Flysch (Table 4.1).

Mt dei Cervi hydrogeological unit

The Mt dei Cervi unit consists of a dolomitic limestone and calcareous-siliceous-marly framework. Various clayey-siltstone units attributable to the Numidian Flysch (pelitic facies) are recognisable in the topography and geomorphology and

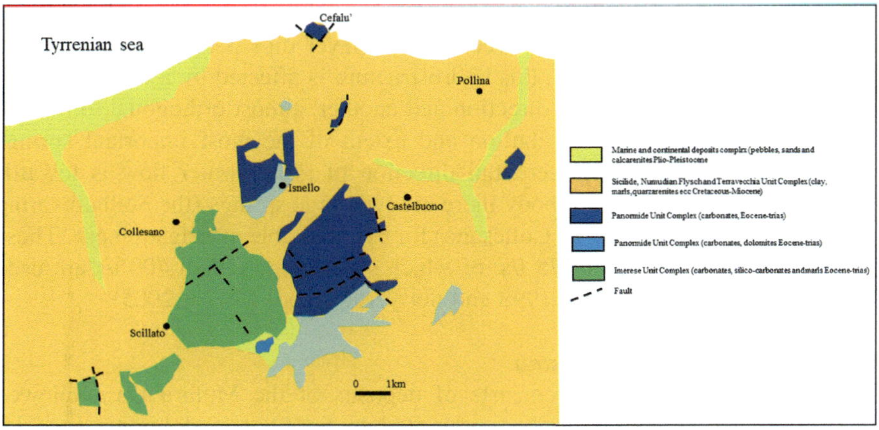

Fig. 4.2 Hydrogeological scheme of the Madonie Mts (modified from Aureli et al. 2001)

Table 4.1 Water bodies and their description

Water body	Hydrogeological complex	Mouton hydrogeological complex	Description	Type of aquifer
Mt dei Cervi	CA Sandstone	Calcareus complex	Calcareous-dolomitic and clacareous-siliceous-marly succession, on which the clayey-marly covers of the Numidian Flysch are based	Unconfined aquifer on fractured and/or karsified limestone
Pizzo Carbonara—Pizzo Dipilo	CA Sandstone	Calcareus complex	Calcareous-dolomitic and clacareous-siliceous-marly succession, on which the clayey-marly covers of the Numidian Flysch are based	Unconfined and confined aquifer on fractured and/or karsified limestone
Mt Quacella	CA/LOC Sandstone Local aquifer	Calcareus complex/Flysch complex	Calcareous-dolomitic and clacareous-siliceous-marly succession, on which the clayey-marly covers of the Numidian Flysch are based	Unconfined and confined aquifer on fractured and/or karsified limestone
Pizzo Caterineci	LOC Local aquifer	Flysch complex	Terrigeous deposits of the Numidian Flysch (Upper Oligocene-low Miocene	Multilayer aquifer

seem to seal the hydrostructure. This can be considered a large anticlinal structure elongated in a north–south direction with layer dips prevailing towards the south and southeast. Moreover, this hydrostructure is affected by a dense network of direct faults with a NE-SW direction and another, almost orthogonal to it, with a NW–SE direction. The distribution and extent of the most important springs seem to indicate that the preferential direction of groundwater flow is towards the SW and NW. In this waterbody there are two large springs (the Scillato spring group and the Favara spring of Collesano) for hydropotable or irrigation use. These sources draw an average of 775 l/s, of which slightly more than 700 l/s are used for drinking water (data up to 1998 and not yet updated) (INGV 2005).

Mt Quacella hydrogeological unit

The stratigraphic succession consists of deposits of the Mufara Un., followed upwards by escarpment deposits represented by infraliassic dolomitic breccias (Fanusi Un.). In places, these dolomites are covered by thin levels of radiolarites (Crisanti Un.) and/or "red scaglia" (Caltavuturo Un.) containing megabreccia carbonate platform elements.

Several springs flow from this hydrostructure, some of which are picked up by the East Madonie Aqueduct (Margi, Margi Soprano, Margi Menta, Gisa, Galleria Cella, Faguara, Romito, Fra Paolo and other springs) that supply the municipalities of Caltanissetta, Enna and minor centres near. Other springs such as Pietà, Urrà, Sorgitore and Grotticelli are managed by the EAS, the Consorzio di Bonifica Palermo 2 and the municipalities of Caltavuturo and Sclafani Bagni.

The distribution and extent of the most important springs seem to indicate that the preferential direction of groundwater flow is towards the SW, SE and NE, respectively towards the Pietà, Urra, Sorgitore and Grotticelli springs group (average flow about 100 l/s), towards the Caterratti springs and towards the Faguara, Romito, Fra Paolo and Canna springs. The springs captured in this water body provide a total flow of just over 550 l/s, of which 370 l/s are for drinking water use (East Madonie Aqueduct), and more than 180 l/s for agricultural use. A significant number of uncaptured springs are present (data updated in 1998).

Pizzo Carbonara-Pizzo Dipilo hydrogeological unit

The Pizzo Carbonara-Pizzo Dipilo hydrogeological unit is made up of mainly carbonate deposits of the central body of the Panormide Carbonate Platform that develop from the Noricum.

The first stratigraphic succession consists of marls, radiolarian calcilutites, resedimented calcarenites, dolomitized breccias (Mufara Un.) of Noric age, in which the Noricum carbonate platform and reef facies continue, followed, after a large hiatus, by soils in back-cliff and reef facies of the Upper Jura—Lower and Middle Cretaceous and by calcilutites with planktonic foraminifera (Scaglia) of the Upper Cretaceous. In discordance with the Jura-Cretaceous deposits, sloped sediments consisting of argillites, marls and biocalcarenites resedimented with large foraminifers, algae and molluscs are found (Gratteri Un.) of Upper Eocene–Oligocene age, followed in turn by arenaceous mudstones and quartzarenites (Numidian Flysch).

The Pizzo Carbonara-Pizzo Dipilo hydrogeological unit is the most suitable geological structure in the Madonie sector for a certain amount of water accumulation, not only because it is essentially made up of intensely karstified and fractured limestone, but also because it is laterally confined by impermeable soils (clays and sandstones of the Numidian Flysch). Pizzo Dipilo also has a generalised groundwater discharge towards the N and NE, but the extent and number of springs are more scarce (Tribuna > 7 l/s, Capo d'Acqua 3.5 l/s). The springs captured in this water body provide a total flow of just over 400 l/s.

Pizzo Catarineci hydrogeological unit

This hydrogeological unit consists of a succession of silty mudstones and quartzarenites (arenaceous facies of the Numidian Flysch) with more or less powerful lenticular intercalations of sandstones and/or quartz conglomerates. This is a multi-strata aquifer with suspended water tables, which are located in the conglomeratic sandstone intercalations. The aquifer is of obvious local hydrogeological interest in the sectors with the greatest number of springs. Permeability is generally discrete and can reach, in some cases, even high average values (up to 10^{-5} m/s), even if, on a large scale, permeability is estimated at between 10^{-6} and 10^{-7} m/s. The aquifer complex rests on the pelitic terms of the Numidian Flysch, which can be considered practically impermeable (10^{-9} m/s).

The degree of vulnerability is very low to medium in the most fractured horizons.

There are several springs in this water body, some of which are captured for drinking water use. The water body includes the Pietra Giordano spring, which is captured and bottled for commercial use (Acqua Geraci).

4.3 Hydrogeochemical Characterisation

In the Madonie hydrological basin 73 springs have been surveyed for potable use (Arpa, PRGA, INGV 2005; Morici et al. 2020). Among these 11 springs have been sampled (Fig. 4.3). The chemical composition of the springs is presented in Table 4.2. Springs discharge at elevations ranging from sea level to 1.355 m a.s.l..

Generally, the temperatures of the springs are negatively correlated with the altitude of emergence and, in consequence, with the relative supply basin. In fact, the Cateratti springs, which emerge around 1,400 m asl, have the lowest temperature (Table 4.2). An exception is shown by the Favara di Collesano spring, which has a higher temperature than excepted for its altitude. This anomaly is probably linked to weak thermal phenomena, since the aquifer feeding this spring is located in correspondence of one of the main tectonic lineaments of the area, probably tapping a deeper aquifer.

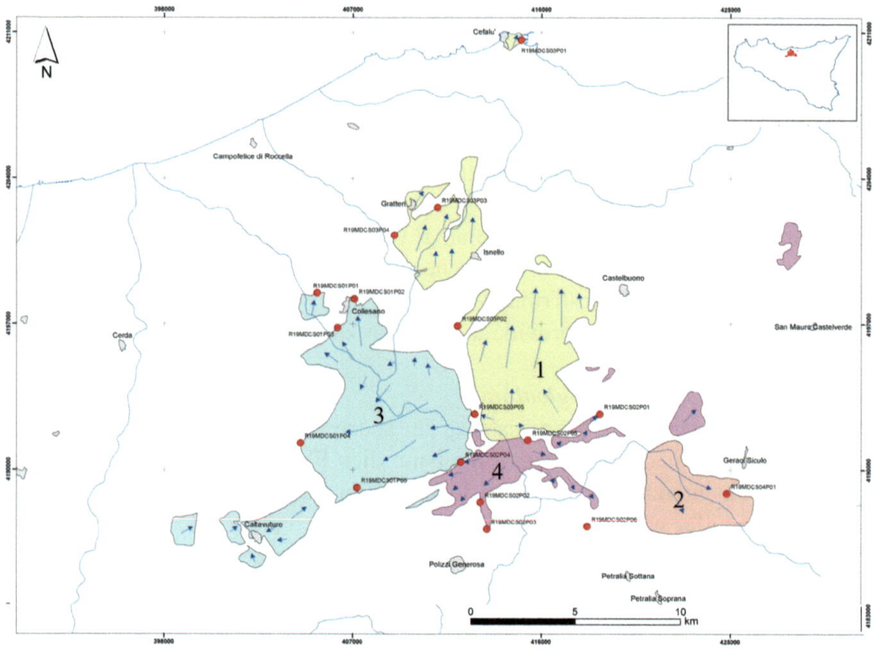

Fig. 4.3 Map of Madonie hydrological basin. Coulors identify bodies: Orange: Pizzo Carbonara-Pizzo Dipilo; light blue Pizzo Catarineci; Magenta: Mt dei Cervi; deep pink: Mt Quacella. Location of spring is shown (green plots represent PRGA springs, blue dots indicate springs sampled for this work)

Both conductivity and pH show values compatible with carbonate aquifers (272 < EC < 531 μ S/cm; 5.7 < pH < 8), except for:

- the Presidiana spring, which emerges at sea level, at the base of the Rocca di Cefalù. This position, also thanks to its feeding system, leads to a weak marine contamination, which, based on the chemical and isotopic composition, has been estimated at around 4–6%.
- the Piano Lana spring, which belongs to the quartzarenitic aquifer of Pizzo Catarineci, with very low salinity (62 μ S/cm).

In the Langelier-Ludwig diagram (Fig. 4.4) most of the waters classify as bicarbonate-alkaline-earth type. These waters represent one of the terms of reference in the geochemical landscape of the Madonie Mountains to which three of the significant water bodies (Mt Carbonara-Pizzo Dipilo, Mt dei Cervi, Mt Quacella) can be ascribed. The other term is represented by the Piano Lana spring, located at a high altitude on the slopes of Mt Catarineci, whose aquifer is set in the quartzarenitic portion of the Numidian Flysch. The chemical composition of

Table 4.2 Chemical and isotopic composition of the sampled springs

	T	EC	pH	Eh	Na	K	Ca	Mg	Cl	NO_3	SO_4	HCO_3	δ^2 H	δ^{18} O	WB
Favara di Collesano	16.8	453	7.5	232	0.51	0.038	2.94	2.1	0.406	0.018	0.6	4.6	−47	−8.31	MC
Scillato	12.8	338	8.00	213	0.29	0.021	2.24	1.45	0.27	0.008	0.34	3.5	−51	−8.76	MC
Canne	9.8	375	7.51	309	0.27	0.018	2.43	1.96	0.258	0.023	0.35	4.1	−51	−8.74	MQ
Grotticelle	10.1	321	7.67	306	0.21	0.017	2.11	1.69	0.172	0.036	0.15	3.7	−53	−9.28	MQ
Cateratti	9.4	405	7.65	261	0.23	0.023	2.77	1.95	0.194	0.011	0.75	4.1	−53	−9.03	MQ
Bevaio del Faggio	8.6	411	7.46	248	0.25	0.02	2.96	1.98	0.177	0.015	0.2	4.85	−52	−8.79	PC-PD
Presidiana	15.6	3587	7.06	225	32.66	0.686	7.6	6.42	38.54	0.047	3.75	5.4	−50	−8.58	PC-PD
San Giorgio	13.6	531	7.24	264	0.48	0.027	4.97	1.15	0.528	0.081	0.26	5.7	−44	−7.72	PC-PD
Capo D'acqua Tribuna	12.7	425	7.52	321	0.23	0.013	4.8	0.18	0.03	0.038	0.15	5.00	−43	−7.67	PC-PD
Favara di Isnello	10.7	273	7.4	331	0.27	0.01	2.62	0.32	0.303	0.067	0.18	2.65	−50	−8.72	PC-PD
Piana Lana	9.6	62	5.68	299	0.25	0.016	0.17	0.13	0.256	0.044	0.15	0.12	−49	−8.72	PC

Analytical data are expressed as meq/L, EC in μ S/cm and δ^2H and δ^{18}O in ‰ versus SMOW. WB is for water body

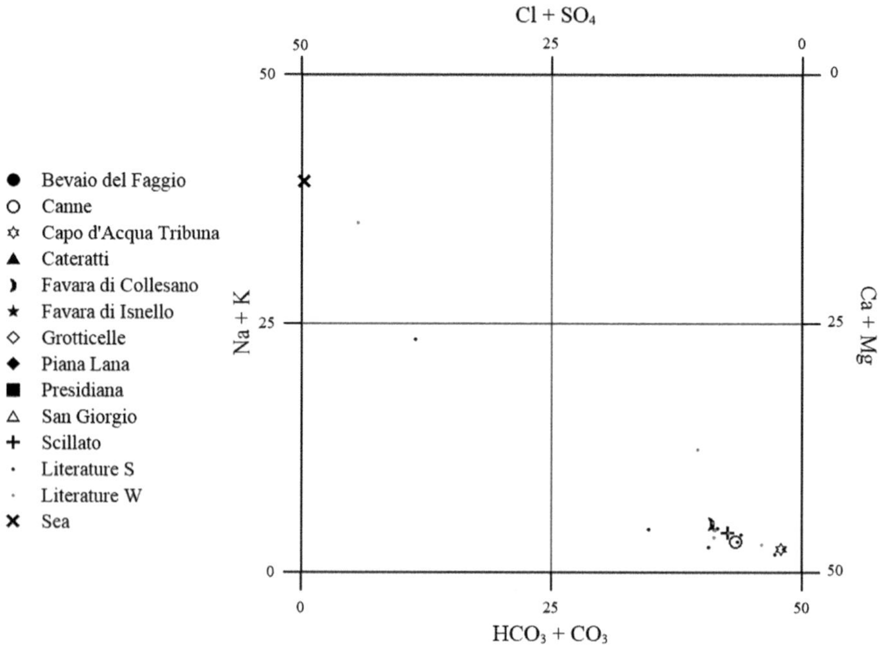

Fig. 4.4 Langelier-Ludwig classification diagram. The spring dataplot in the bicarbonate-alkaline-earthy, Chlorine-sulphate-alkaline-earthy and Chlorine-sulphate-alkaline quadrants. Literature data are taken from Morici et al. (2020)

this water falls in the chlorine-sulfate-alkaline-earth quadrant, and is characterized by a low salt content.

The Presidiana spring falls in the chlorine-sulfate-alkaline water quadrant because of the marine contamination it undergoes (4–6%). This type of contamination compromises the usability of this water for drink water purposes. Currently, in order to use it, it is mixed with low saline water from Scillato spring.

The triangular diagrams, related to the concentrations of the main cationic and anionic species (Fig. 4.5), highlight HCO3 and Ca as prevalent ions, in agreement with the prevalent calcareous composition of the aquifer rocks. The triangular diagram of anionic species (Fig. 4.5 left), confirms the differentiations already identified with the Langelier-Ludwig diagram (Fig. 4.4), i.e. a prevalence of bicarbonate waters, with the exception of Piano Lana and Presidiana. The calcium/magnesium ratio (Fig. 4.5 right) allows to differentiate the water springs associated to purely calcareous with those belonging to dolomitic aquifers. To the first group belong sources such as Favara di Isnello and San Giorgio, which insist on soils constituted by the most markedly calcareous terms of the Panormide Platform, which are opposed by emergencies such as Canna and Cataratti, characterized by a Ca/Mg ratio referable to more dolomitic aquifers (Dolomites of Mt. Mufara).

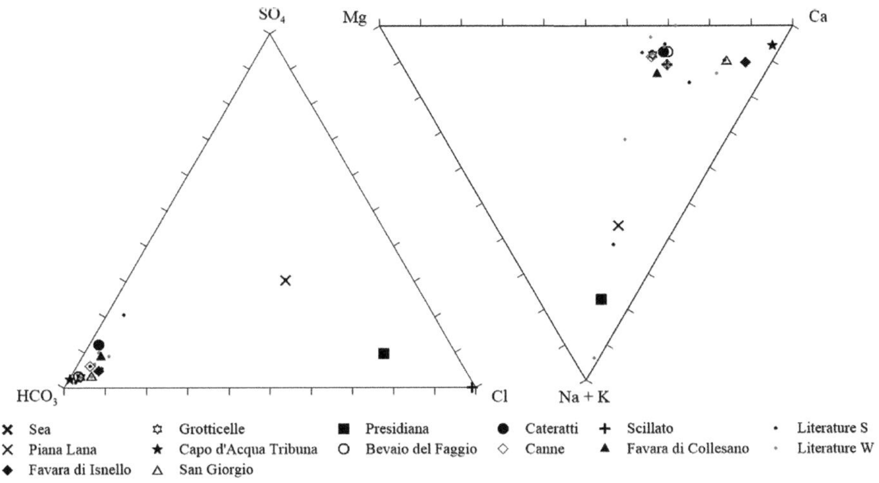

Fig. 4.5 Anions (left) and cations (right) ternary diagrams for Madonie Mountains catchment area. It is evident the Presidiana seawater pollution. Literature data are shown (INGV 2005). S: Spring, W: Wells

4.4 Isotopic Characteristics of the Catchment Area

The waters taken in the Madonie hydrogeological basin intercept the MMWL line of the meteoric waters of the Mediterranean Sea (Fig. 4.6).

The LMWL for Madonie basin was determined by INGV et al. (2005) and Liotta et al. (2013) as sector 1 (see Cap. 1, Table 1.6). The INGV LMWL ($\delta^2H = 6.3\delta^{18}O + 4.92$) and the Liotta LMWL ($\delta^2H = 6{,}31\delta^{18}O + 5$) are practically coincident and overlapping.

The sampled and literature waters show a certain compositional homogeneity in both δ^2H and $\delta^{18}O$ ($-53 < \delta^2H < -43$; $-9 < \delta^{18}O < -8$ ‰). The Presidiana spring shows a slight ($\delta^2H = -50$‰; $\delta^{18}O = -8.59$‰) compared to the other springs due to the marine input it undergoes. Using the equation relative to sector 1 of Table 1.7, it is possible to go back to the zone of average supply assessable over 1500 m above sea level compatible both in terms of altitude and in terms of geological-structural and hydrogeological characteristics with the Madonie mountain chains (Fig. 4.7).

4.5 Trace Elements

Trace elements in Madonie basin, show a relative ample range variation per element (Table 4.3, Fig. 4.8). The highest concentration in almost all trace elements are in Presidiana sample because of the long distance water goes through runs from the feeding altitude (> 1500 m a.s.l.) up to the sea level where it springs (Table 4.3).

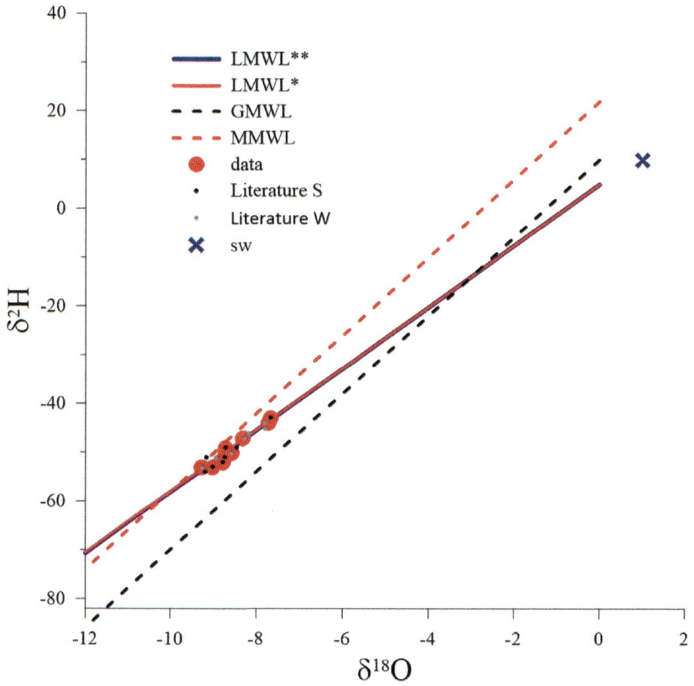

Fig. 4.6 δ^2H—$\delta^{18}O$ (in ‰ respect to SMOW) diagram of spring waters. The LMWL* is from INGV (2005), LMWL** is from Liotta et al. (2013). The GMWL and MMWL are shown for comparison. Literature data from INGV (2005). S: Spring, W: Wells

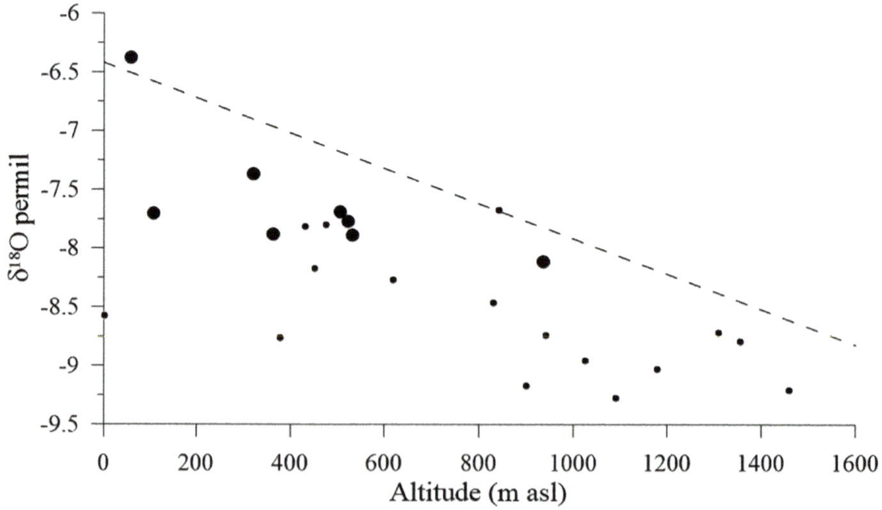

Fig. 4.7 Relationship between $\delta^{18}O$ and altitude

Table 4.3 Trace elements in sampled waters

	Li	B	Al	V	Cr	Mn	Fe	Co	Ni	Cu	Zn	As	Se	Rb	Sr	Sb	Ba	Pb	U	WB
Favara di Collesano	5.82	51.3	0.82	0.22	bdl	bdl	10.3	bdl	1.07	0.25	0.39	0.35	2.38	0.945	552	bdl	44.8	bdl	0.586	MC
Scillato	3.09	27.0	0.6	0.14	bdl	bdl	2.00	0.10	2.64	0.26	1.04	0.17	0.68	0.454	325	bdl	26.3	0.083	0.372	MC
Canne	2.18	18.6	1.3	0.41	0.14	bdl	10.5	bdl	0.89	0.27	0.89	4.48	0.35	0.455	114	bdl	200	bdl	0.487	MQ
Grotticelli	0.75	8.1	2.46	0.38	0.19	bdl	8.44	bdl	0.65	0.29	0.4	2.31	0.28	0.186	74.8	bdl	42.3	bdl	0.247	MQ
Cateratti	2.64	20.5	1.00	bdl	bdl	bdl	13.5	bdl	1.09	0.15	0.38	7.36	0.54	0.478	220	0.05	85.6	bdl	1.21	MQ
Presidiana	bdl	348.5	1.15	0.55	5.55	0.15	316.15	bdl	3.55	6.7	3.00	29.6	45.75	nd	741	0.13	46.3	0.05	0.5	PC-PD
Favara di Isnello	0.4	15.9	1.41	0.33	0.25	bdl	9.98	bdl	1.01	0.18	0.25	0.65	0.18	0.314	52.1	bdl	21.6	bdl	0.13	PC-PD
Capo d'Acqua Tribuna	0.3	11.4	0.6	0.66	0.5	bdl	22.1	0.12	1.48	0.18	0.3	7.59	0.51	0.264	110	bdl	57.2	bdl	0.259	PC-PD
San Giorgio	1.61	16.4	0.36	0.26	0.25	bdl	21.9	0.11	1.55	0.16	0.31	8.24	0.59	0.463	95.7	bdl	243	bdl	0.172	PC-PD
Bevaio del Faggio	2.76	18.9	0.72	0.45	bdl	0.11	10.8	bdl	1.31	0.41	0.18	0.55	0.17	0.319	265	bdl	49.4	bdl	0.342	PC-PD
Piano Lana	1.94	11.3	9.82	bdl	bdl	4.88	5.44	0.15	2.12	0.18	4.7	0.21	0.17	0.74	23.0	bdl	55.0	bdl	bdl	PC

Concentrations are in μ g/L. WB is for water body. MC is Mt dei Cervi, MQ is Mt Quacella, PC-PD is Pizzo Carbonara-Pizzo Dipilo and PC is Pizzo Catarineci

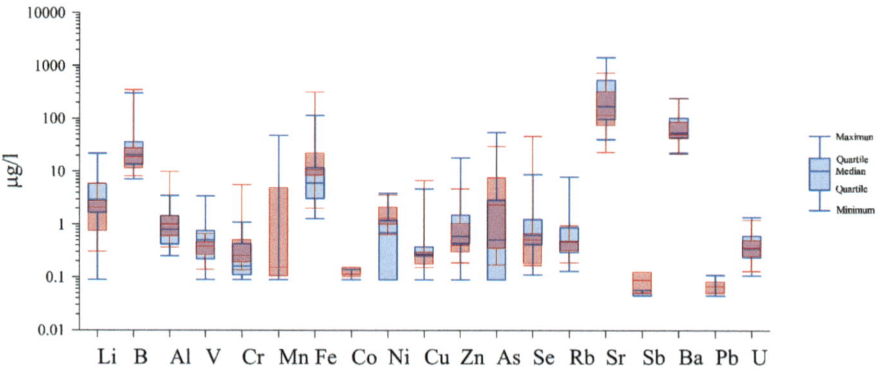

Fig. 4.8 Box-whisker plot of trace element at Madonie basin. Red box spring data (this work). Light blue box data from Morici et al (2020)

Concentrations of trace elements in sampled spring and literature are always below the threshold of the legal limits (Fig. 4.8).

Bibliography and Recommended Further Reading

Abate B, Pescatore T, Renda P, Tramutoli M (1988a) Geological scheme of the Termini Imerese Mountains and the western Madonie Mountains. Mem Soc Geol It 41:465–474

Abate B, Renda P, Tramutoli M (1988b) Illustrative notes of the geological map of the termini Imerese Mountains and the Western Madonie Mountains (north-central Sicily). Mem Soc Geol It 41:475–505

Agnesi V, De Cristoforo D, Di Maggio C, Macaluso T, Madonia G, Messana V, Rotigliano E (1998) Introduction to the geomorphology of the Madonie Mountains. 79th National Congress of the Italian Geological Society, Palermo

Alaimo R, Carapezza M, Dongarrà G, Hauser S (1987) Geochemistry of sicilian thermal springs. In: "Thermal resources of Sicily and minor islands." Institute of Mineralogy, Petrography and Geochemistry, University of Palermo. Sicilian Region, Department of Land and Environment., 47–65, STASS, Palermo

Aureli A, Contino A, Cusimano G, Silluzio C (2001) Groundwater vulnerability of the Madonie Mountains. 7th Conference on Limestone Hydrology and Fissered Media, Besançon (France), 20–22 September 2001, Département de Géosciences, Université de Franche-Comté, Besançon

Aureli A, Contino A, Cusimano G, Silluzio C (2002) Map of pollution vulnerability of Madonie aquifers—(scale 1:50.000). C.N.R., SELCA, Firenze

Badalamenti F, Coco B, Dongarrà G, Hauser S, Parello F (1982) Isotopic hydrology of Sicily. The Springs of Cefalù. Rend. SIMP 38:795–802

Carapezza M, Cusimano G, Liguori V, Alaimo R, Dongarrà G, Hauser S (1987) Introductory note to the study of thermal springs on the Island of Sicily. In: Thermal resources of Sicily and minor islands. Institute of Mineralogy, Petrography and Geochemistry, University of Palermo, Sicilian Region, Department of Land and Environment., 7–46, STASS, Palermo

Coltro R, Ferrara V (1980) The water resources of the Rock of Cefalù in the hydrogeological framework of north-central Sicily. In: IV international conference on groundwater, III international conference on water planning. Capomulini Unni Acireale (Catania), 17–21 February 1980, vol 3, pp 149–176

Craig H (1961) Isotopic variations in meteoric waters. Science 133:1702–1703

Cusimano G (1987) Study and estimation of groundwater resources of the Madonie territory for the purpose of its geological-environmental assessment. In: Ecological Culture Seminar, Cycle IV: Interdisciplinary analysis of the Madonie Territory, 17 March 1978, Castelbuono (Palermo)

Cusimano G, Contino A (1995) Hydrogeological map of the Madonie Park (scale 1:25,000). Madonie Park territorial plan, Madonie Park Authority—Sicily Region (internal report)

Cusimano G, Favara R, Francofonte S, Madonia P, Valenza M (1992) Hydrogeological and hydrogeochemical features of the Madonie carbonate mountain range (Sicily). In: Alpine Caves: Alpine Karst system and their environmental context, International Congress, Asiago (VI), 11–14 June, pp 207–214

Favara R, Dongarrà G, Hauser S, Longinelli A (1984) Geochemical-isotopic study of a series of springs in the Scillato area (Palermo). Rend Soc It Min Petr 39:421–427

Gat JR, Carmi I (1970) Evolution of isotopic of atmospheric waters in the Mediterranean sea area. J. Geophis. Res. 75:1437–1440

Grasso M, Lentini F, Vezzani L (1978) Stratigraphic-structural lineaments of the Madonie Mountains (north-central Sicily). Geol Romana 17:45–69

Istituto Nazionale di Geofisica e Vulcanologia (INGV), Sezione di Palermo (2005) First characterization and monitoring of groundwater aimed at the drafting of the "Water Protection Plan of the Sicily Region". On behalf of the Sicilian Region—Deputy Commissioner for the waste emergency and water protection

Liotta M, Grassa F, D'Alessandro W, Favara R, Gagliano Candela E, Pisciotta A, Scaletta C (2013) Isotopic composition of precipitation and groundwater in Sicily, Italy. Appl Geoch 34:199–206

Morici S, La Pica L, Favara R, Gagliano Candela E, Pecoraino G, Scaletta C (2020) Chemical-physical parameters, major, minor and trace elements composition of groundwater belonging to the hydrogeological basin of the Madonie Mountains in Sicily (Italy). Interdisciplinary Earth Data Alliance (IEDA). https://doi.org/10.26022/IEDA/111732

Ministry LL.PP.-Hydrographic Service (1934) The Italian springs. Sicilia, vol II, Palermo Hydrographic Service. Polygraphic Institute dello Stato, Rome

5 Nebrodi Mountains Hydrogeological Basin

Leonardo La Pica, Sabina Morici, Esterina Gagliano Candela, and Giovannella Pecoraino

5.1 Geological Framework

The tectonic units recognisable in the Nebrodi territory (Fig. 5.1) occupy the geometrically highest positions within the Sicilian mountainous structure.

Numidian successions constitute the Maragone Unit (deepest), the Mt Salici-Mt Castelli Unit and the Nicosia Unit (highest). The latter two represent terrigenous bodies deposited on a Sicilide substrate (Fig. 5.2).

The Sicilide successions also constitute three main tectonic units, which, starting from the deepest, are the Mt Soro Unit, the Upper Flaky Clays Unit and the Troina Unit.

The compressive deformation of this sector of the Sicilian chain begins in the Upper Oligocene—Lower Miocene and continues at least until the Upper Pliocene, as demonstrated by the presence of evaporitic deposits and of the folded and faulted Trubi in the Mt Sambughetti area.

The geometric relationships of the tectonic units constituting the framework of the Nebrodi and Peloritani Mountains can be summarised as follows:

L. La Pica · S. Morici · E. Gagliano Candela (Deceased) · G. Pecoraino (✉)
Istituto Nazionale di Geofisica e Vulcanologia, Sezione di Palermo, Via Ugo La Malfa 153, 90146 Palermo, Italy
e-mail: giovannella.pecoraino@ingv.it

L. La Pica
e-mail: leonardo.lapica@ingv.it

S. Morici
e-mail: sabina.morici@ingv.it

E. Gagliano Candela et al. (eds.), *Drinking Water in Sicily*,
https://doi.org/10.1007/978-3-031-60473-7_5

Fig. 5.1 The Nebrodi Mountains. Imagine from Google Earth Pro v 7.3.6.9345 (December 29, 2022). Sicily, Italy

Late Pre-collisional Covers of the Lower-Middle Miocene

- *Reitano Flysch (Upper Oligocene?-Langhiano)*

  Sicilide Units

- *Troina Unit*
- *Nicosia Unit (Numidian Flysch)*
- *Upper flaky clays Unit*
- *Mt Soro Unit*
- *Mt Salici-Mt Castelli Unit (Numidian Flysch)*

  Panormide Units

- *Maragone Unit (Numidian Flysch)*

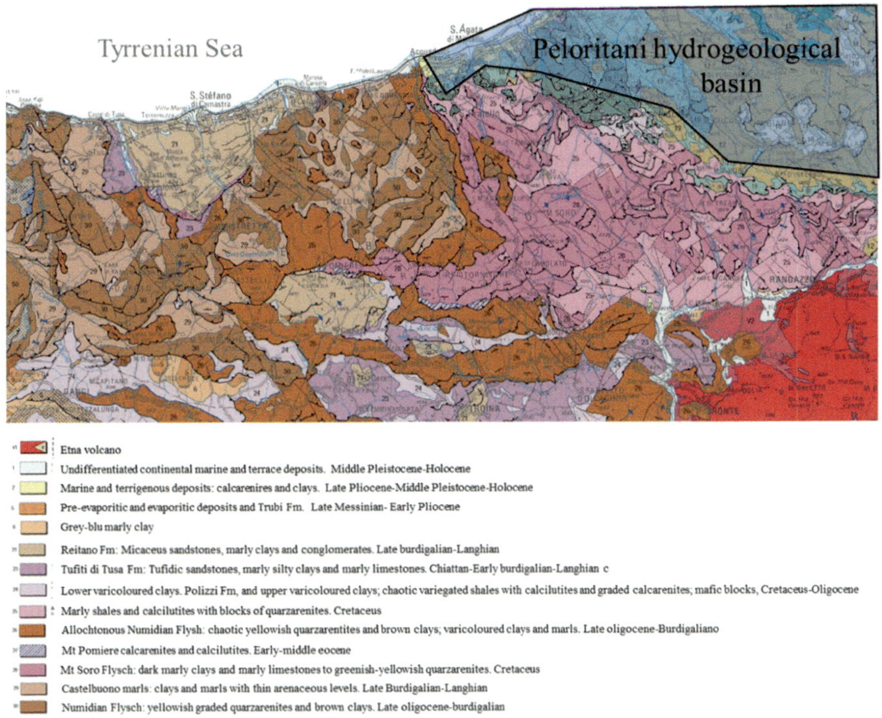

Fig. 5.2 Geological sketch map of the Nebrodi Mountains (modified from Lentini et al. 2014)

5.1.1 Late Pre-Collisional Cover of the Lower-Middle Miocene

Reitano Flysch (Upper Oligocene?-Langhiano).
It appears as a clayey-arenaceous-conglomeratic alternation from 500 to over 800 m thick that covers, in discordance, the Sicilide tectonic units. The Reitano Flysch succession is composed, from bottom to top, of calcilutites and conglomerates passing to lithic sandstones (calciclastic turbidites), clays, marls and grey sandstones with a carbonate matrix that, towards the top, pass to dark grey marls to nummulites. This portion of the succession is replaced by a sequence several hundred metres thick (>500 m) represented by an alternation of grey-yellowish micaceous sandstones, sometimes graded and rather coarse, poorly cemented, in layers from 20 cm to 2 m, and grey-greenish marly clays in layers that are generally decimetric in thickness. From a petrographic point of view, these are generally immature arkosic sandstones, composed of fragments of feldspar, quartz, volcanites and phyllites, with spatially variable compositions.

5.1.2 Sicilidae Unit

Troina Unit
This unit is mainly found in the sectors of the Ancipa basin. It consists of a basal portion of red and green Varicoloured Clays, with frequent intercalations of micaceous siltstones, sandstones and metamorphic clasts and calcarenites (mixed internal facies in Nigro and Renda, 2000). This interval has a considerable thickness and evolves stratigraphically upwards to the Troina-Tusa Flysch succession, i.e. the Miocene flysch successions, including the Reitano Flysch.

Nicosia Unit (Numidian Flysch)
This unit occurs in association with the pelitic sequences of the Sicilidae Complex. The base of the unit consists of dark grey Varicoloured Clays. The pelitic intervals, with frequent exotic arenaceous inclusions, possess considerable thickness, especially in the outcropping sectors south of S. Teodoro. This interval is replaced towards the stratigraphic top by Varicoloured Clays with a prevalent reddish colouring, very similar in facies to those of the Superior Scaly Clays. This interval evolves upwards to brown clays and quartzarenites very similar to those constituting the Numidian Flysch. The unit overlies the Mt Salici-Mt Castelli Unit and constitutes, as a whole, a southward dipping monoclinal structure within which there are several tectonic repetitions of the succession.

Upper Flaky Clay Units
This unit is always found at the top of the Mt Soro Unit and consists of a chaotic lithosome with a prevalent grey clayey matrix containing exotic blocks referable to different terms of the underlying unit. The Upper Flaky Clay constitutes a continuous horizon on which the overthrust of the Peloritani Units has occurred. They are highly unstable and almost impermeable.

The Varicoloured Flaky Clays are represented by marly clays and dark grey marls, alternating in places with grey marly limestones in centimetric layers and graded calcarenites. Upwards, polychrome mudstones and arkosic-lithic sandstones are present.

Mt Soro Unit
The Mt Soro Unit geometrically overlies the Mt Salici-Mt Castelli Unit, with an exposed contact along the S. Fratello-Cesarò alignment. It constitutes among others the reliefs of Portella Femmina Morta, Portella Miraglia, Mt Soro and Pizzo Antenna. To the south it extends as far as Mt Abate (Cesarò), while to the east it emerges as far as Serra del Re. To the west, it outcrops as far as Poggio Tornitore and Mt Pomiere.

The unit is constituted of basal clayey-carbonate horizons evolving towards the stratigraphic top to an argillitic-quartzarenitic interval until evolving to an apical interval constituted by quartzarenitic banks. The succession ends with grey argillites with thin quartzarenitic intercalations. The basal horizons consist of an alternation of dark grey flaky marly clays and brown marly limestones, the latter pervaded by a network of fractures filled with spatic calcite.

The clayey-carbonate succession is made up of alternating blackish-grey calcilutites and/or calcarenites, with thicknesses varying from 10 to 20 cm, and grey foliated argillites, with thicknesses varying from 20 to 50 cm (Mt Soro Flysch, carbonate member, Cretaceous).

In the clayey-arenaceous interval, the quartzarenites are medium-grained, without evident gradation and are greenish-yellow in colour. The sandstones have a sub-arkosic composition overall (Carmisciano et al. 1983). The succession consists of alternating greenish or dark grey coarse quartzarenites and green or wine-red foliated argillites, with bank strengths of 0.8–1 m (Mt Soro Flysch, quartzarenitic member, Palaeogene-Cretaceous). The thickness of the Mt Soro Flysch exceeds 300 m.

Mt Salici-Mt Castelli Unit (Numidian Flysch)
The unit represents the geometrically deepest tectonic unit in the segment of the Sicilian chain outcropping in the Nebrodi Mountains. It has a thickness of at least 600 m, with a lower clay interval, which sometimes has grey marls at its base. The quartzarenite beds often have a lenticular geometry. The unit outcrops extensively from Castel di Lucio up to the height of the Furiano Stream.

5.1.3 Panormide Units

Maragone Unit (Numidian Flysch)
It occurs extensively in the Nebrodi Mountains and probably constitutes the terrigenous cover of the Panormide Carbonate Platform or its margin. It is composed of quartzarenites evolving upwards to an alternation of silty clays and quartzarenitic banks up to 50 m thick. This portion of the Numidian succession is stratigraphically replaced by a *thinning* and *fining upward* sequence of silicoclastic turbidites, which upward are replaced by a predominantly clayey interval. The soils referable to this tectonic unit constitute the Nebrodic ridge.

The numidic successions are represented by an alternation of brown or blackish siliceous argillites and siltstones and grey quartzarenites and quartzosiltstones, sometimes in large banks. The argillitic portions contain limonitic nodules in places and vary in thickness from a few decimeters to a few meters. The quartzarenites vary in thickness from a few decimeters to several tens of meters. They are made up of banks of sandstones of an average thickness of 2–2.5 m, alternating with layers of pelites of an average thickness of 15–25 cm.

The sandstones, ochre-yellow or brown in colour, consist mainly of quartz granules, while the pelites, varying in colour from dark green to brown, are extremely foliated and can therefore be considered as argillites.

The successions of the Numidian Flysch of the Maragone Unit are organised in several overlapping minor-order tectonic bodies with a southern vergence, which together form a tectonic stack several thousand metres thick.

5.2 Hydrogeology

The water bodies of the Nebrodi Mountains hydrological basin have different hydraulic characteristics and resources, depending on the degree and type of permeability of the rocky media and their geometric characteristics (hydrostructures). The permeability of outcropping soils varies in relation to both the primary (porosity) and secondary (cracking) characteristics, the latter resulting from the degree of tectonisation to a greater or lesser extent achieved by the rocky bodies (Ferrara 1972, 1973).

Terrains with different permeabilities can be distinguished:

- *very high permeability due to primary porosity*: alluvial deposits at the bottom of the valley of streams and torrents, dejection conoids and particularly thick detrital prisms;
- *high permeability due to cracking permeability*: to this category belong the conglomeratic facies of the Reitano Formation. Where they are permeated by fault networks and neotectonic fractures, permeability can become very high;
- *medium–high permeability due to porosity and/or cracking*: the proximal arenaceous facies of the Reitano Formation;
- *average permeability due to cracking and/or porosity*: numidic successions where quartzarenitic intercalations are thicker;
- *medium–low permeability due to porosity and/or cracking*: the distal clayey-arenaceous alternation of the Reitano Formation, the numidic successions where the quartzarenitic intercalations are less thick, and to a lesser extent the sicilid quartzarenites;
- *very low permeability*: variegated clays, numidic and sicilidi clay successions.

In the Nebrodi Mountains hydrogeological basin groundwater circulation is poorly developed, with the exception of the water bodies constituted by the Reitano Flysch successions, which represent hydrogeological reservoirs of discrete potential.

Groundwater flow within the Numidian successions is limited and localised within the main neotectonic dislocation zones, while the Sicilidian clay successions represent an impermeable horizon. The lithologies with the highest permeability are mainly alluvial deposits (recent and actual alluvium), concentrated along the narrow coastal plains and along the main rivers and quaternary deposits consisting mainly of gravels, sands and silts often associated with unconfined aquifers. The remaining part of the basin (over 90%) is characterised by medium–low permeability associated with flyschoid lithotypes.

The following water bodies outcrop in the Nebrodi Mountains (Fig. 5.3):

- Reitano-Mt Castellaci
- Capizzi-Portella Cerasa
- Pizzo Michele-Mt Castelli
- Mt Ambola

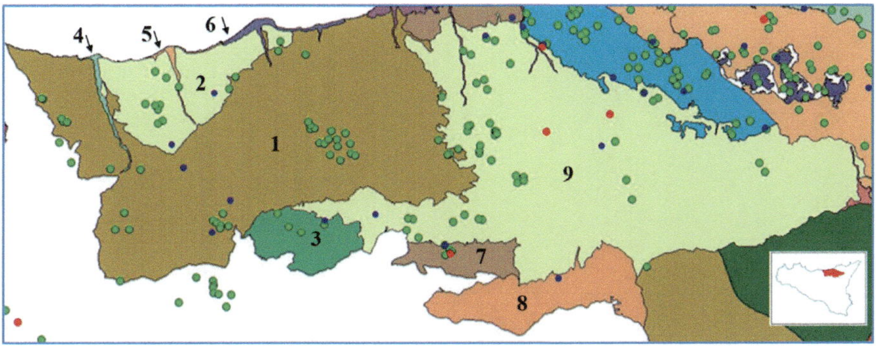

Fig. 5.3 Map of Nebrodi hydrological basin. Numbers identify water bodies: (1) Pizzo Michele-Mt Castelli; (2) Reitano-Mt Castellasi; (3) Capizzi-Portella Cerasa; (4) Tusa; (5) S. Stefano; (6) Caronia; (7) Mt Abola; (8) Cesarò-Mt Scalonazzo; (9) Mt Soro. Location of springs is shown (green plots represent PRGA springs, INGV, 2005, the blue dots indicate springs sampled for this study)

- Cesarò-Mt Scalonazzo
- Tusa fluvial-coastal aquifer
- S. Stefano fluvial-coastal aquifer
- Caronia fluvial-coastal aquifer
- Mt Soro

Among them, Mt Ambola and Cesarò-Scalonazzo water bodies are not of hydrogeological interest and will not be considered or described. In Table 5.1 the description of each water body is presented.

5.3 Hydrogeochemical Characterisation

In the Nebrodi Mountains hydrogeological basin, nine water bodies have been identified. Of these, two (Monte Ambola and Cesarò-Monte Scalonazzo) are not of hydrogeological interest. In the remaining seven water bodies (Table 5.2, Fig. 5.3), four new samples were taken. At each site, T, conductivity and pH were determined in the field and other basic parameters were determined in the laboratory. The temperature varied between 7 and 14 °C, the conductivity between 62 and 905 S/cm (at 20 °C), and the pH varied between 6.1 and 8.2 (Table 5.2).

The geochemical characterisation of groundwater is conditioned by the presence of flyschoid lithotypes (Flysch Numidico, Flysch di Monte Soro and Flysch di Reitano) and alluvial soils deposited along the rivers (Tusa, Santo Stefano and Caronia). The waters circulating in the flyschoid lithotypes present a relatively low content of calcium and magnesium and hardness commonly lower than 150 mg/L $CaCO_3$, while the waters connected to the alluvial aquifers are characterised by a higher content of Ca and Mg, and hardness higher than 150 mg/L $CaCO_3$.

Table 5.1 The description of the Nebrodi Mts water bodies

Water body	Hydrogeological complex	Mouton hydrogeological complex	Description	PTA hydrogeological complex	Type of aquifer
Tusa	Valley flooding	Complex of alluvial deposits of major rivers	The alluvium of the Fiumara di Tusa consists mainly of gravel and blocks with frequent small and thick plastic silty lenses in the medium–low section	Recent deposits	Free aquifer
Santo Stefano	Valley flooding	Complex of alluvial deposits of major rivers	The water body coincides with the alluvial aquifer of the Fiumara di Santo Stefano, which develops longitudinally in a NNO-SSE direction	Recent deposits	Free aquifer
Caronia	Valley flooding	Complex of alluvial deposits of major rivers	The Caronia torrent consists mainly of quartzarenitic blocks, which are associated with gravel and pebbles in the lower and middle stretches, with frequent plastic silty lenses	Recent deposits	Free aquifer
Reitani-Mt Castellaci	Local groundwater	Flysch complex	Turbiditic arenaceous-clayey succession (Reitano Flysch) of Upper Burdigalian-Serravallian age	Clay-silica-marnaceous flyscioid complexes of Permo-Triassic-Oligomiocene age	Multilayer aquifer
Pizzo Michele-Mt Castelli	Local groundwater	Flysch complex	Regular monoclines characterised by quartzarenitic banks alternating with brown clays of Lower Miocene age resting on a substratum of greyish clays with varicoloured sections of Upper Oligocene age	Clay-silica-marnaceous flyscioid complexes of Permo-Triassic-Oligomiocene age	Local confined water tables

(continued)

Table 5.1 (continued)

Water body	Hydrogeological complex	Mouton hydrogeological complex	Description	PTA hydrogeological complex	Type of aquifer
Mt Soro	Local groundwater	Flysch complex	Alternations of coarse greenish or dark grey quartzarenites and green or burgundy-red foliated argillites (Monte Soro Flysch, quartzarenitic member); alternations of blackish-grey calcilutites and/or calcarenites and grey foliated argillites (Monte Soro Flysch, carbonate member; Upper Flaky Clays)	Clay-silica-marnaceous flyscioid complexes of Permo-Triassic-Oligomiocene age	Local confined water tables
Capizzi-Portella Cerasa	Local groundwater	Flysch complex	The water body is located in a turbiditic arenaceous-clayey succession (Reitano Flysch) of Upper Burdigalian-Serravallian age	Clay-silica-marnaceous flyscioid complexes of Permo-Triassic-Oligomiocene age	Local confined water tables

Table 5.2 Chemical and isotopic composition of the sampled waters

	T	EC	pH	Eh	Na	K	Ca	Mg
Pomiere-Mercadante	11.1	401	7.76	177	0.33	0.014	4.00	0.70
Fossa Neve	9.4	172	7.26	262	0.48	0.023	1.24	0.31
Priola	11.1	369	8.21	182	0.51	0.043	2.63	1.28
Cicè	13.1	635	8.40	170	1.10	0.047	4.47	2.67
	Cl	NO_3	SO_4	HCO_3	δ^2H	$\delta^{18}O$	WB	
Pomiere-Mercadante	0.34	0.016	0.17	4.46	− 47	− 8.85	CPC	
Fossa Neve	0.22	0.008	0.2	1.56	− 49	− 8.48	MS	
Priola	0.46	0.082	0.49	3.39	− 39	− 7.28	MS	
Cicè	0.62	0.05	0.66	6.9	− 45	− 7.71	RMC	

The low salinity that characterises the waters of this basin is linked to the low water–rock interaction, caused by the poorly soluble chemical species of the flyschoid lithotypes, and to a relatively shallow circulation.

Analytical data are expressed as meq/L, EC in μS/cm and δ^2H and $\delta^{18}O$ (in ‰ versus SMOW). WB stands for water body, CPC for Capizzi-Portella Cerasa, MS for Mt Soro, RMC for Reitano-Mt Castellaci water bodies.

In the Langelier–Ludwig diagram (Fig. 5.4), the newly sampled spring waters and those from literature (INGV, 2005) fall in the range of bicarbonate-alkaline-earthy, chlorinated-sulfate-alkaline-earthy and alkaline chlorinated sulfate fields. The evolution of the waters from earthy alkaline to alkaline could be related to the water interactions with flyschoid lithologies (Flysch of Mt Soro, Numidian Flysch and Flysch of Reitano), where the clay component is considerable. The clay minerals that form these sediments, in contact with natural waters, tend to adsorb bivalent ions (Ca and Mg) with respect to monovalent ions (Na and K), enriching the latter (Na preferentially to K) in the aqueous solution. Instead, waters richer in Ca and Mg came in contact with carbonate soils or interacted with carbonate cement present in the arenitic component of flyschoid sequences.

For this reason, in the ternary diagram of the cations (Fig. 5.5 right) starting from the calcium vertex, two trends can be seen: one towards a component richer in magnesium (interaction with the carbonate cement) and another towards one richer in alkali (interaction with the flysch clays). Moreover, in the anion ternary diagram (Fig. 5.5 left), two trends can be recognized, starting from the HCO_3 vertex, towards chlorides and sulphates. Waters enriched in sulfates are not the same as waters that are enriched in chlorides. In this case, the trends are not related to the same type of interaction described for cations. It is necessary to imagine the presence of different lithologies (one richer in sulfate and the other richer in chlorides) such as the Messinian chalks present in small patches in the basin or from the oxidation of hydrogen sulfide produced by the degradation of organic matter contained within the clay deposits of the flyschoid sequences. Direct interactions between aquifers and seawater are excluded. The presence of Cl

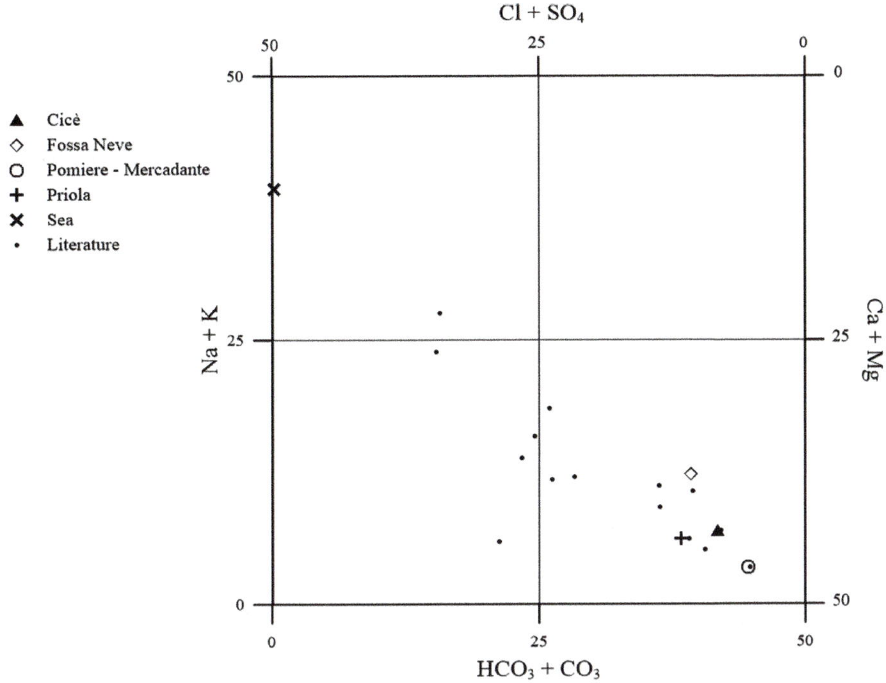

Fig. 5.4 Langelier–Ludwig classification diagram. The springs plot in the bicarbonate-alkaline-earthy quadrant. Literature data (INGV, 2005) are shown for comparison; S: spring, W: well

depends entirely on marine aerosols in stormwater precipitations, which constitute groundwater recharge.

5.4 Isotopic Characteristics of the Catchment Area

The hydrogeological basin of the Nebrodi Mountains is characterized by a dorsal ridge (up to 1.500 m high), which develops in E-W direction. The hydrogeological basin of Peloritani Mountains, adjacent to it to the East, is characterized by a ridge of about 1.000 m high, which stretches in a SW–NW direction. This leads to the distinction of two zones with different climatic characteristics: one along the coastal lines, with a thermal regime typical of Mediterranean areas, whereas, the other one is that of the mountain ridge, where the climate tends to become more continental. This characteristic affects the isotopic composition of precipitation and hence of groundwater. The isotopic composition of groundwater varies from −9‰, in high altitude areas, to −7‰ in coastal areas for $\delta^{18}O$, and from −49 to −39‰ for $\delta^{2}H$ (Table 5.2). The waters taken in the Nebrodi Mountains hydrogeological basin align on and very close to the MMWL (Fig. 5.6). The LMWL for the Nebrodi and Peloritani basins was determined by Liotta et al. (2013) as

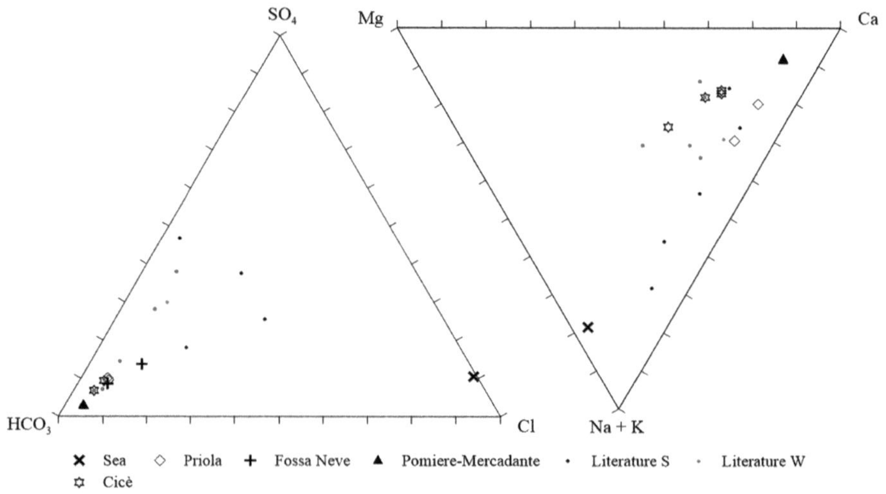

Fig. 5.5 Anions (left) and cations (right) ternary diagrams for Nebrodi Mountains catchment area. Literature data are shown (INGV 2005). S: spring, W: well

sector 3 (see Chap. 1, Table 1.6). The LMWL is defined as $\delta^2H = 6.14\ \delta^{18}O + 3.9$. Collected and literature water data fall close to the LMWL, confirming the meteoric origin of these waters.

The isotopic composition of springs is characterized by rather more negative values than the local meteoric waters (INGV, 2005) suggesting a considerable contribution from feeding areas site at higher altitudes (Fig. 5.7) and therefore a groundwater circulation that connects groundwater and water bodies located at high altitudes with those located at lower altitudes.

According to the isotopic gradient of 0.13δ‰/100 m (Table 1.7 relative to sector 3, Nebrodi and Peloritani), the average altitude of supply is around 1.500 m a.sl., compatible with the average altitude of the basin.

5.5 Trace Elements

Like in the other hydrogeological basins, trace element concentrations in spring waters are highly variable (Fig. 5.8; Table 5.3), but always below the legal limits (Table 1.1) for drinking waters.

Cicè spring, in the Reitano-Mt Castellaci water body, has the highest concentration of Li, B, Ba and Sr, probably due to the dissolution of the carbonate matrix in marls and sandstones.

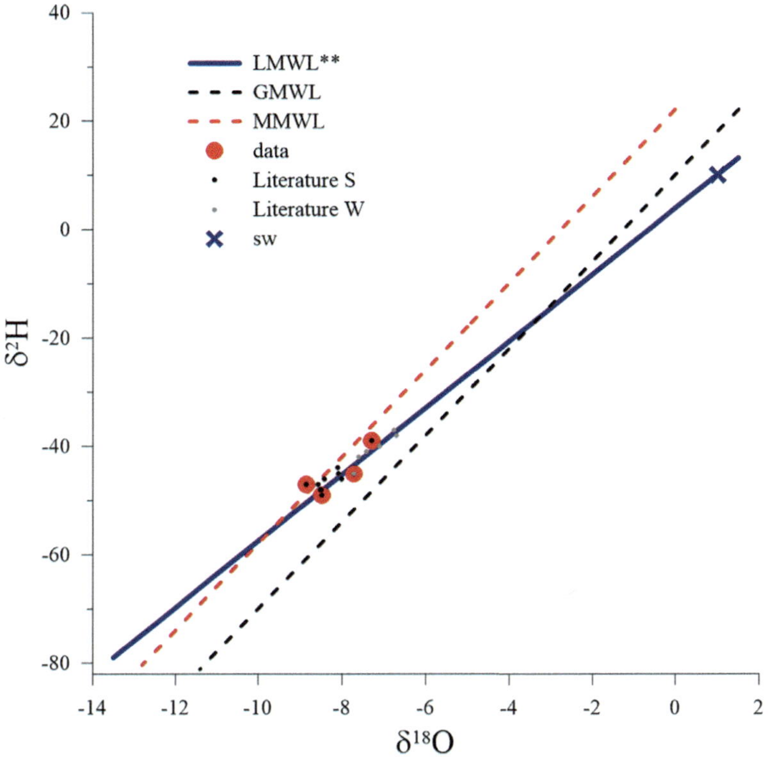

Fig. 5.6 δ^2H–$\delta^{18}O$ (‰ versus SMOW) diagram of spring waters. LMWL** is from Liotta et al. (2013). The GMWL and MMWL are shown for comparison. Literature data from INGV (2005). S: spring, W: well

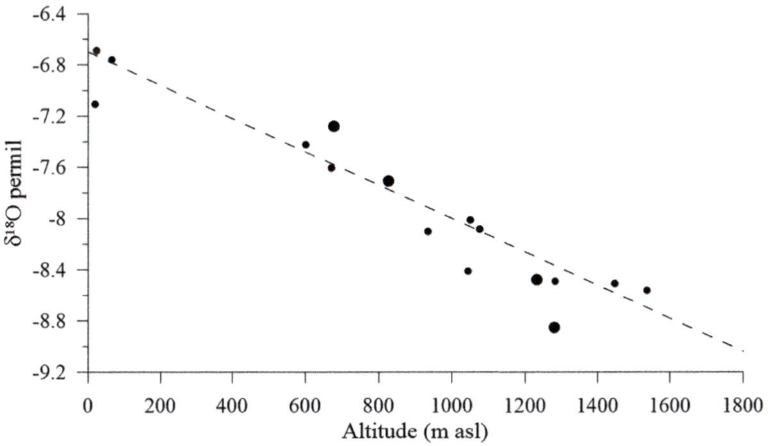

Fig. 5.7 Relationship between $\delta^{18}O$ and altitude

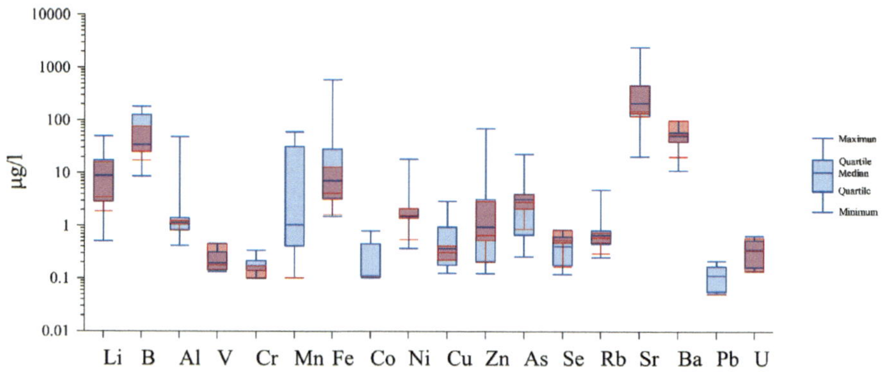

Fig. 5.8 Box-whisker plot of trace elements at Nebrodi Mountains basin. Red box spring data are results from this study; light blue box are data from INGV (2005)

Table 5.3 Trace elements in sampled waters

	Li	B	Al	V	Cr	Mn	Fe	Co	Ni	Cu
Pomiere	1.88	17.5	1.26	0.18	0.17	bdl	13.1	bdl	1.43	0.23
Fossa Neve	2.88	25.3	1.08	bdl	bdl	bdl	1.7	bdl	0.55	bdl
Priola	4.13	26.3	0.83	0.46	0.1	bdl	3.2	bdl	1.75	0.42
Cicè	16.16	75.8	1.09	0.15	0.14	0.1	5.1	0.11	2.14	0.32

	Zn	As	Se	Rb	Sr	Ba	Pb	U	WB
Pomiere	2.96	4.00	0.49	0.302	164	20.7	0.052	0.14	CPC
Fossa Neve	0.21	0.88	0.17	0.485	121	40.2	bdl	bdl	MS
Priola	0.81	2.15	0.85	0.759	138	67.8	bdl	0.351	MS
Cicè	0.53	3.58	0.59	0.72	470	101.0	bdl	0.590	RMC

Concentrations are in µg/L. CPC is for Capizzi-Portella Cerasa, MS is Mt Soro, RMC is Reitano-Mt Castellaci water bodies

Bibliography and Recommended Further Reading

Accordi B (1958) Report on the geological surveys of 1956 in the Nebrodi Mountains (northeastern Sicily). Boll Serv Geol It 77(1):167–174

Baldacci L (1886) Geological description of the Island of Sicily. Mem Descr Della Carta Geol 1:408

Carmisciano R, Pennisi M, Puglisi D, Romeo M (1983) Preliminary data on the composition, provenance, and age of calcareous turbiditic levels of the M. Soro Flysch (Nebrodi Mountains, north-central Sicily). Boll Acc Gioenia Sc Nat Catania 16(321):13–129

Duée G (1962) The limestone series underlying the Reitano formation in the Caronia-Mistretta-Castel di Tusa region, Nebrodi mountains (northern Sicily). Boll Soc Géol France 4(7):728–733

Duée G (1965) Paleogeography of the Nebrodi Mountains (Sicily). Bull Soc Géol France 7(7):889–899
Duée G (1969) Geological study of the Nebrodi Mountains (Sicily). Thése Fac Sc Paris, p 424
Duée G (1970) The geology of the Nebrodi Mountains of Sicily. In: Geology and history of Sicily. Petrol. Expl. Soc. Libya, pp 187–200
Ferrara V (1972) Hydrogeological map of Sheet 611 "Mistretta," scale 1:50.000. Serv Geol d'It, Cartrographic Editorial Graphics, Rome
Ferrara V (1973) Presentation of the hydrogeological map at the scale 1:50,000 of the Mistretta Sheet. In: Proceedings 2nd international Conv on Subterranean Waters, 28 Apr 2 Mag 1973, Palermo, pp 401–423
General Regulatory Plan of Sicilian Aqueducts. 2006–2012–2017
Istituto Nazionale di Geofisica e Vulcanologia (INGV), Sezione di Palermo (2005) First characterization and monitoring of underground waters aimed at the drafting of the "water protection plan of the region of Sicily". On behalf of the Sicilian Region-Commissioner Delegate for the Waste Emergency and Water Protection
Lentini F, Carbone S (2014) Geology of Sicily. Mem Descr Carta Geol d'It. XCV, pp 7–414
Lentini F, Catalano S, Carbone S (2000) Geological map of the province of Messina. Regional Province of Messina, Tip. Selca, Firenze
Liotta M, Grassa F, D'Alessandro W, Favara R, Gagliano Candela E, Pisciotta A, Scaletta C (2013) Isotopic composition of precipitation and groundwater in Sicily, Italy. Appl Geoch 34:199–206
Nigro F, Renda P (2000) A model of tectono-sedimentary evolution of the Sicilian Neogene foreland. Boll Soc Geol It 119:667–686
Ogniben L (1960) Explanatory note on the geological scheme of northeastern Sicily. Riv Min Sic 2:183–212
Schmidt di Friedberg P (1963–1964) Petroleum lithostratigraphy of Sicily. Riv Min Sic 88–90:198–217
Vezzani L (1972) Geological map of Italy 1:50,000, sheet 611, Mistretta Ser Geol It, Roma
Wezel FC (1970) Geology of the numidian flysch of northeastern Sicily. Mem Soc Geol It 9:225–280

Hydrogeological Basin of the Palermo Mountains

6

Esterina Gagliano Candela, Leonardo La Pica, Giovannella Pecoraino, and Sabina Morici

6.1 Geological Framework

The hydrogeological basin of the Palermo Mountains includes a sector of the province of Palermo and a large part of the western and eastern portion of the Palermo Metropolitan Area (Fig. 6.1). The Oreto River, with its valley rising into the mountain range, divides it into two large sectors: the south-eastern sector, which includes the whole watershed with the mountains of Piana degli Albanesi and the branches towards the Jato and Eleuterio Rivers and the coast east of Palermo; the north-western sector, which includes the western half-arc of the mountains facing the Conca d'Oro, with the western extensions facing the Gulf of Castellammare.

The Palermo Mountains constitute a segment of the northern Sicilian chain and are the result of the tectonic superimposition of various carbonate, carbonate-silicoclastic and terrigenous geological bodies of the Upper Triassic–Lower Miocene, generally with a north–south orientation. These bodies derive from Miocene deformation, both of a pelagic basin and escarpment succession (Domain Imerese Auct.), and of carbonate platform successions (Panormide Domain and Trapanese Domain Auct.) identified during the Mesozoic distensive phases and surrounded by silicoclastic terrigenous deposits (Numidian Flysch Auct.).

E. Gagliano Candela (Deceased) · L. La Pica · G. Pecoraino (✉) · S. Morici
Sezione di Palermo, Istituto Nazionale di Geofisica e Vulcanologia, Via Ugo La Malfa 153, 90146 Palermo, Italy
e-mail: giovannella.pecoraino@ingv.it

L. La Pica
e-mail: leonardo.lapica@ingv.it

S. Morici
e-mail: sabina.morici@ingv.it

E. Gagliano Candela et al. (eds.), *Drinking Water in Sicily*,
https://doi.org/10.1007/978-3-031-60473-7_6

Fig. 6.1 The Palermo Mountains. Imagine from Google Earth Pro v 7.3.6.9345 (December 29, 2022). Sicily, Italy

In the middle to upper Pliocene, further deformation of the tectonic edifice is associated with NW–SE transverse fault systems, which resulted in the extrusion of the geometrically lower tectonic bodies into the structural edifice, producing retrovergent out-of-sequence systems. A second NE-SW system, with a transcurrent component, cuts off the previous one and is responsible for some hydrothermal vents.

The structural analysis of the entire tectonic edifice of the Palermo Mountains has allowed us to recognise a wedge of carbonate platform units (Panormide and Trapanese units) covered by basin units (Imerese units) that constitute the highest part of the tectonic edifice (Fig. 6.2). There are also open fractures with a N-S trend, which play an important hydrogeological role by inducing the development of the hypogeum karst network and conditioning the underground water flow in the hydrostructures. The geometric relations of the tectonic units constituting the structure of the Palermo Mountains can be summarised as follows:

- Trapanese Unit
    - Mt Kumeta Unit
- Panormidae Units
    - Mt Pecoraro-Pizzo Vuturo-Mt Pellegrino Unit
    - Mt Palmeto–Mt Castellaccio Unit
    - Mt Gallo Unit

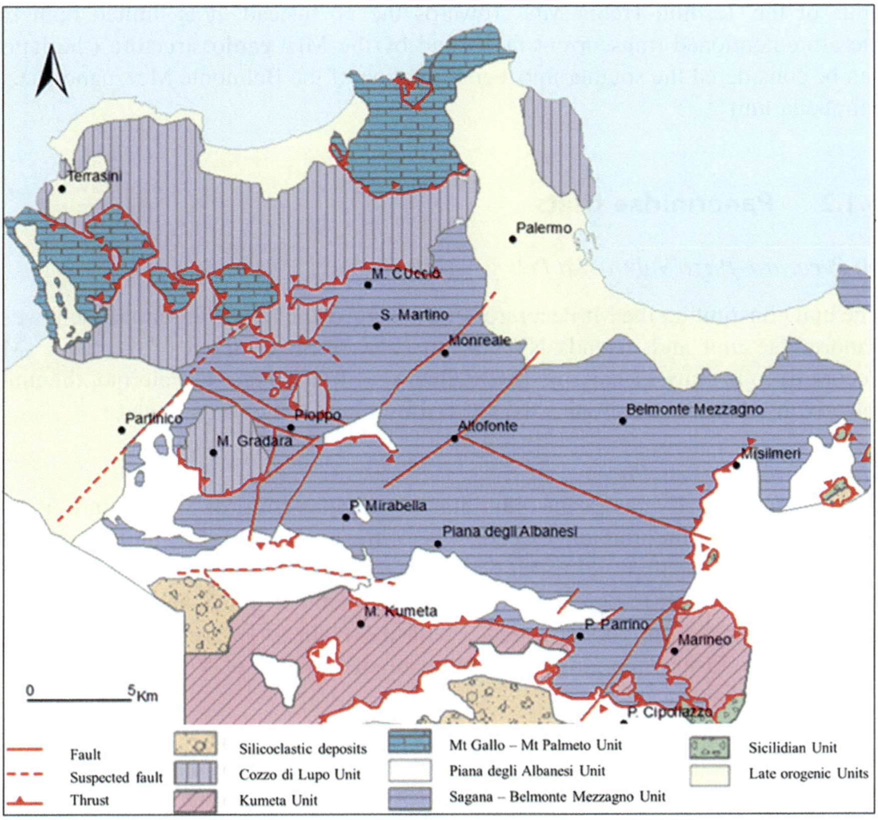

Fig. 6.2 Geological scheme of the Palermo Mountains (modified from Abate et al. 1978)

- Imerese Units
  - Belmonte Mezzagno Pizzo Mirabella Unit
  - Mt Gradara Unit
  - Mt Saraceno-Mt Cuccio Unit
- Sicilidi Units
- Late and postorogenic deposits

6.1.1 Trapanese Units

Mt Kumeta Unit

This unit represents the lowest unit in the structural edifice and is dissected by E-W transverse faults, rotating eastwards to NW–SE, that have determined the extrusion of the relief. It emerges uninterruptedly from S. Giuseppe Iato–S. Cipirello up to the surroundings of Marineo and Bolognetta where it is covered by the Imeresi units of the Termini-Trabia Mts. Towards the N, instead, it is limited both by the aforementioned transcurrent faults and by the Mt Leardo structure (the latter can be considered the southernmost continuation of the Belmonte Mezzagno-Pizzo Mirabella unit).

6.1.2 Panormidae Units

Mt Pecoraro–Pizzo Vuturo–Mt Pellegrino Unit

The unit constituting the Mt Pecoraro, Pizzo Vuturo and Mt Pellegrino is the lowest Panormidae unit and extends NW–SE to E-W from the town of Terrasini (Mt Pecoraro) to the city of Palermo (Mt Pellegrino). In the Plain of Palermo, the unit, lowered by distensive tectonics, is covered by Pleistocene calcarenites.

Mt Palmeto–Mt Castellaccio Unit
This unit tectonically covers the Numidian terrigenous cover of Mt Pecoraro–Pizzo Vuturo Mt Pellegrino unit and extends NW of the city of Palermo until it includes, in addition to the main relief of Mt Castellaccio, also the Isola delle Femmine. The overlapping front on the previous unit, clearly visible to the S and SE, is instead masked to the east, in correspondence with the Plain of Palermo, by Pleistocene calcarenite deposits, emerging only in the abraded mound of Villa Raffo.

Mt Gallo Unit
This represents the highest unit among the Panormidae units that constitute the tectonic edifice of the Palermo Mountains and is dissected by both reverse faults and direct to transtensive faults. The overlapping front on the previous unit, in correspondence with the Plain, is masked by Pleistocene marine and continental deposits. The Panormidae Units overlap with the Imerian ones, whose thickness reaches up to 1.500 m.

6.1.3 Imerese Units

Belmonte Mezzagno–Pizzo Mirabella Unit

This unit is composed by an anticlinal ridge, repeatedly dissected by extensional and/or transcurrent tectonics. As a whole, the unit consists of a ramp anticline verging towards the S, a buried syncline filled by Numidian terrigenous deposits, partially dissected by distensional tectonics, and the Mt Leardo structure that directly overlies the Trapanese units.

Near the front of the ramp anticline there are calcareous or calcareous-silicicoclastic plates, which can be interpreted as the product of collapses and/or the sliding of blocks or megablocks (e.g. Cozzo Lavanche) and to which modest springs are linked. In correspondence with the northern portion overlooking the Plain of Palermo (Ciaculli, Falsomiele, Villabate, etc.) the anticlinal structure is dissected by direct faults plunging towards the sea with rejections often exceeding 1.000 m and is responsible for springs.

Mt Gradara Unit
This unit consists of a monoclinal structure that overlaps with the previous one and is, like the latter, dissected by distensive tectonics with direct NNE-SSW faults cut by direct ENE-WSW faults and by left-handed transcurrent faults running from NE-SW (Monreale-Pioppo fault) to N-S rotating to NNE-SSW (Torretta fault).

Mt Saraceno–Mt Cuccio Unit
This is the highest of the Imerese units constituting the tectonic structure of the Palermo Mountains. It is formed by both large and totally isolated calcareous-dolomitic plates (megaklippen of Mt Saraceno, Pizzo Montanello and Pizzo Peluso), and by flakes embricked with Numidian Flysch (Mt Fior dell'Occhio, Mt S. Isidoro) and large flaked bodies (Mt Cuccio), which only locally involve the terrigenous cover. In correspondence with the Palermo Plain, the unit is most probably covered, in discordance, by Pleistocene calcarenitic deposits.

6.1.4 Sicilidi Units

These units consist of more or less extensive plates, mostly brachisinclinal, of Varicolored Clay Auct. and overlying marly limestones and calcareous marls with biocalcarenitic or biocalciruditic intercalations (Fm. di Polizzi Generosa Auct.) resting tectonically on the Numidian Flysch.

6.1.5 Late- and Post-orogenic Deposits

To the south-west of the tectonic build-up of the Palermo Mountains, some brachyssynclines are visible in which tardorogenic soils outcrop, mainly composed of pelitic-arenaceous-conglomeratic terrigenous deposits of the Serravallian-Tortonian. In the Palermo Plain there are Pleistocene deposits consisting of blue clay sands and calcarenites and/or well-cemented pebbly conglomerates.

6.2 Hydrogeology

The area of the Palermo Mountains is interesting from a hydrogeological point of view. The relatively high values of annual rainfall and the particular existing hydrogeological structures make this area an important sector of central-northern Sicily.

In this area the different lithological units refer to two hydrogeological complexes:

(a) dolomitic calcareous complex, constituting the main aquifers of the Palermo Mountains, with very low primary porosity and medium-high permeability due to fracturing and karstification, locally well developed;
(b) a substantially impermeable complex of clays, argillites, radiolarytes and marls, which may contain aquifers with ephemeral water tables, or occasionally incorporate permeable bodies of variable dimensions (calcareous megabreccia).

Various lithological terms with a similar degree of permeability have been grouped together in the dolomitic limestone complex, justifying their association. These are the calcilutites and calcarenites of the Scillato Auct. Unit of the upper Trias, the dolomitic breccias, dolorudites and doloarenites of the Fanusi Auct. Unit of the lower Lias, and the large resedimented carbonate bodies (essentially calcarenites and calcirudites) of the Crisanti Auct. Due to its hydrogeological characteristics, it represents one of the main aquifers of the Palermo Mountains, as attested by the presence of some important springs (e.g. Gabriele). The Fanusi dolomite and vacuolar dolomitic breccias and breccias present an indistinct or poorly visible stratification.

Vulnerability is considered high to very high, depending on the greater or lesser karstification of the limestone. The prevailing permeability is high due to fracturing and, in places, porosity, when the complex is intensely karstified ("floury" dolomite of Baida and Cozzo S. Isidoro) and the karst forms are little developed compared to other hydrogeological complexes.

The numerous units pertaining to the Sicilidi and Oligomiocene Units are united in the impermeable complex due to their lithological and hydrogeological affinities. In fact, the following are represented in this complex: the Serravallian–Tortonian clays (Pre-Terravecchia Fm Auct.), the marls and sandy clays of the Tortonian (Terravecchia Unit Auct.), the thinly stratified clays, siltstones and sandstones of the Numidian Flysch (pelitic facies), the Varicoloured Fm Auct. Clays, frequently tectonised, sometimes chaotic, and incorporating a more or less abundant stone fraction. These various composite terms perform the function of an impermeable. The Numidian Flysch presents slightly different characteristics in its pelitic facies due to the location of the arenaceous levels incorporated in the argillites, which are sometimes intensely tectonised. The relative permeability of this complex varies greatly from level to level and from zone to zone, but in any case it remains remarkably low. Permeability is, therefore, low to very low with values between 10^{-7} and 10^{-9} m/s.

Groundwater circulation in the dolomitic limestone complex only occurs through networks of fractures and karst conduits. It is also conditioned by the lithostratigraphic characteristics of the successions and the geometry of the overthrust fronts, which constitute limits and thresholds of permeability of the springs. The relative relationships between the clay coverings (Numidian Flysch) and the carbonate successions define the geometries of the hydrostructures, some of which are hydraulically autonomous (Fig. 6.3). The geometry of drainage systems, within the hydrostructures, is made more articulated by the interaction of several families

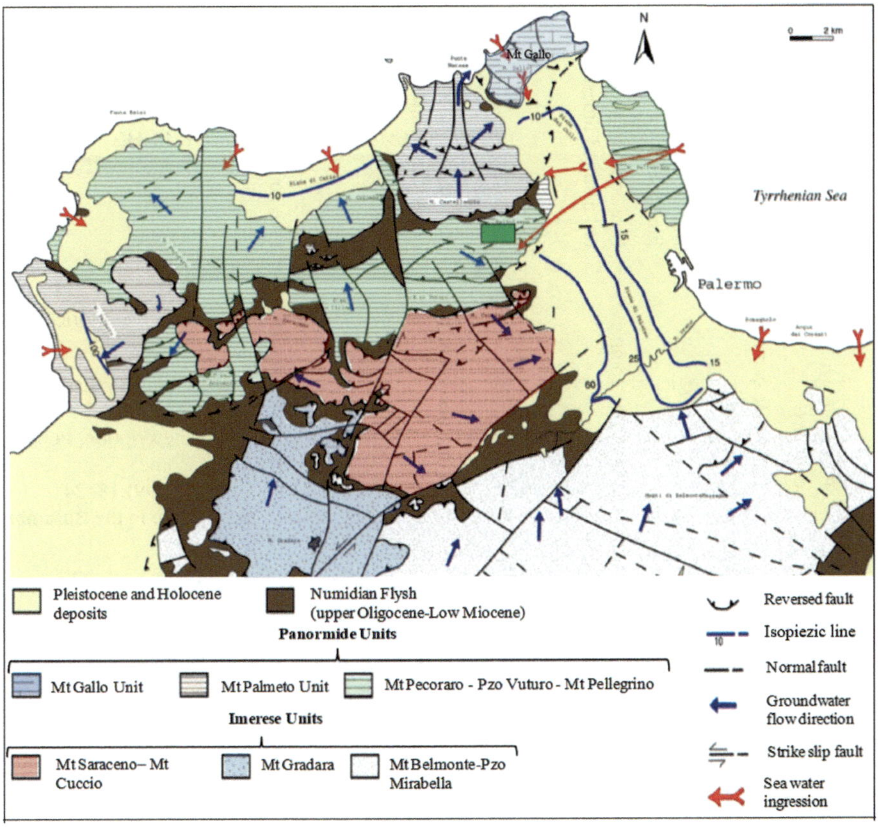

Fig. 6.3 Hydrostructural model of Palermo Mts (modified from Contino et al. 1998)

of fractures and their density, shape and size, as well as by the possible presence of occluding materials and the development of karstification processes.

Within the Mountains of Palermo hydrogeological basin, seven hydrogeological units can be distinguished, some of which consist of several hydrostructures, corresponding to water bodies. A total of eleven hydrostructures, have been identified.

The main hydrogeological units and their hydrostructures (water bodies) are listed below (Table 6.1 and Fig. 6.4):

Mt Kumeta Hydrogeological Unit

- *Mt Kumeta hydrostructure (Mt Kumeta water body)*: This hydrostructure constitutes the ridge with a E-W trend and represents the southern border of the mountains of Palermo. It presents an approximately N to S groundwater flow. The *Traversa* spring is located in this hydrostructure.

Table 6.1 Water bodies and their description

Water body	Hydrogeological complex	Mouton hydrogeological complex	Description	PTA hydrogeological complex	Type of aquifer
Belmonte-Pizzo Mirabella	CA limestone	Calcareous-dolomitic complex and calcareous complex	Carbonate and carbonate-silicoclastic succession	Limestone complex	Aquifer is both free and confined with possible suspended aquifers in fractured and/or karstified limestones
Mt Castellaccio	CA limestone	Calcareous-dolomitic complex	Succession of calcareous-dolomitic soils resulting from the deformation of the Panormide carbonate platform	Limestone complex	Free aquifer fractured and/or karstified limestones
Mt Pecoraro	CA limestone	Calcareous-dolomitic complex	Succession of calcareous-dolomitic soils resulting from the deformation of the Panormide carbonate platform	Limestone complex	Free aquifer fractured and/or karstified limestones
Mt Saraceno	CA limestone	Calcareous-dolomitic complex	Succession of calcareous-dolomitic soils resulting from the deformation of the Panormide carbonate platform	Limestone complex	Free aquifer fractured and/or karstified limestones

(continued)

Table 6.1 (continued)

Water body	Hydrogeological complex	Mouton hydrogeological complex	Description	PTA hydrogeological complex	Type of aquifer
Mt Cuccio-Mt Gibilmesi	CA limestone	Calcareous-dolomitic complex	Succession of calcareous-dolomitic soils resulting from the deformation of the Panormide carbonate platform	Limestone complex	Free aquifer fractured and/or karstified limestones
Pizzo Vuturo-Mt Pellegrino	CA limestone	Calcareous-dolomitic complex	Succession of calcareous-dolomitic soils resulting from the deformation of the Panormide carbonate platform	Limestone complex	Free aquifer fractured and/or karstified limestones
Mt Kumeta	CA limestone	Calcareous-dolomitic complex	Succession of calcareous-dolomitic soils resulting from the deformation of the Panormide carbonate platform	Limestone complex	Free aquifer fractured and/or karstified limestones
Mt Mirto	CA limestone	Calcareous-dolomitic complex	Succession of calcareous-dolomitic soils resulting from the deformation of the Panormide carbonate platform	Limestone complex	Free aquifer fractured and/or karstified limestones

(continued)

Table 6.1 (continued)

Water body	Hydrogeological complex	Mouton hydrogeological complex	Description	PTA hydrogeological complex	Type of aquifer
Mt Gradara	CA limestone	Calcareous-dolomitic complex	Succession of calcareous-dolomitic soils resulting from the deformation of the Panormide carbonate platform	Limestone complex	Free and confined aquifer fractured and/or karstified limestones
Mt Palmeto	CA limestone	Calcareous-dolomitic complex	Succession of calcareous-dolomitic soils resulting from the deformation of the Panormide carbonate platform	Limestone complex	Free aquifer in karst network

Analytical data are expressed as meq/L, EC in μS/cm and δ^2H and $\delta^{18}O$ in ‰ versus SMOW. WB is for water body. BMP is for Belmonte Mezzagno–P.zzo Mirabella, MC is Mt Castellaccio, MCMG is for Mt Cuccio–Mt Gibilmesi, MG is for Mt Gradara, MM is for Mt Mirto, MP is for Mt Pecoraro, MK is Mt Kumeta water bodies

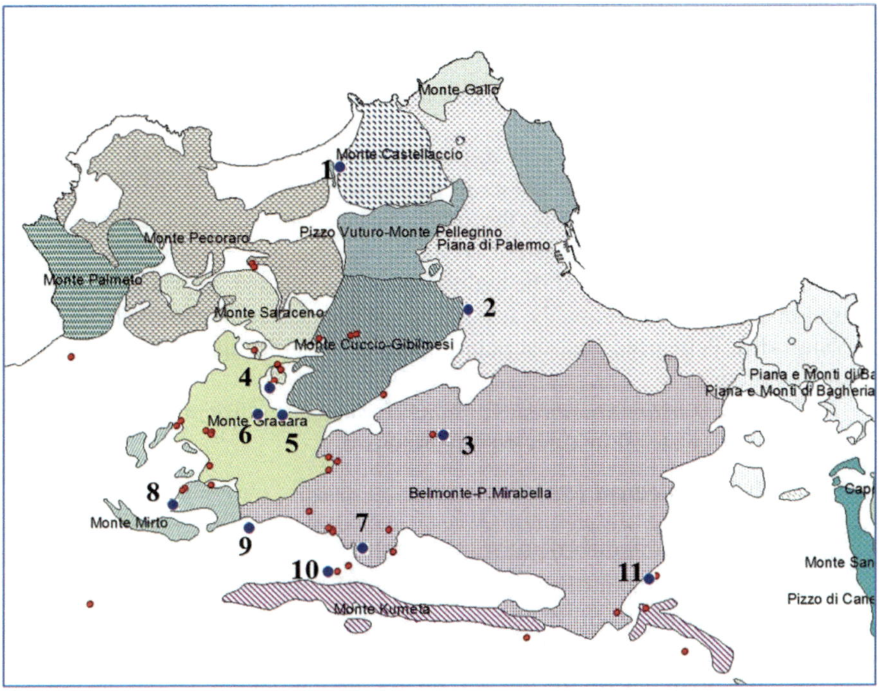

Fig. 6.4 Hydrological basin of Palermo Mts. Different colours indicate water bodies. Location of spring is shown (green plots represent PRGA springs, INGV, 2005, bleu plots spring sampled for this work). (1) is S. Rosalia; (2) Gabriele; (3) S. Maria; (4) Giardinello; (5) Locogrande; (6) Soprasansotta; (7) Frassino; (8) Mirto; (9) Grisì; (10) Kumeta; (11) Risalaimi

Mt Pecoraro-Pizzo Vuturo-Mt Pellegrino Hydrogeological Unit

- *Pizzo Vuturo-Mt Pellegrino hydrostructure (Pizzo Vuturo-Mt Pellegrino water body):* two main components of the subterranean water flow towards N and NE can be assumed, which in the direction of Mt. Pellegrino rotates towards the east. This hydrostructure is affected by a wedge of marine intrusion that extends over 7 km from the coastline and also affects the calcarenitic aquifer of the Plain of Palermo. The carbonate aquifer of Pizzo Vuturo continues below the Mt Castellaccio aquifer with which, however, it does not seem to be hydraulically connected, since the two aquifers are separated by disconnected strips of shale of the Numidian Flysch, and the component of the N-direction groundwater flow moves under the Mt Castellaccio aquifer and discharges into the sea beyond Punta Matese.
- *Mt Pecoraro hydrostructure (Mt Pecoraro water body):* the Mt Pecoraro hydrostructure is affected by a bundle of direct, N-S trending faults, the main one acting as a subterranean watershed. The water flow is directed both towards

the Piana di Carini and the Punta Raisi sector. The *Belvedere alta* spring is located in this hydrostructure.

Mt Palmeto-Mt Castellaccio Hydrogeological Unit

- *Mt Castellaccio hydrostructure (Mt Castellaccio water body):* in this hydrostructure the groundwater flow moves in different directions, the main component towards the Piana dei Colli, the others respectively towards the sea (Punta Matese) and towards the eastern sector of the Piana di Carini (Fig. 6.3). The south-easternmost sector of the hydrostructure is affected by groundwater salinisation phenomena, probably related to marine ingression through the Pizzo Vuturo-Mt Pellegrino hydrostructure. The *S. Rosalia* spring belongs to this hydrostructure.
  - *Mt Palmeto hydrostructure (Mt Palmeto water body:* It constitutes a megaklippe with water flow directed SW, seaward.

Mt Gallo Hydrogeological Unit

- *Mt Gallo hydrostructure (Mt Gallo water body)* presents a trend of the water flow both towards the Plain of the Hills and towards the sea and does not present any kind of spring manifestations.

Belmonte Mezzagno-Pizzo Mirabella Hydrogeological Unit

- Mt Mirto hydrostructure (Mt Mirto water body): this hydrostructure includes the Mt Signora relief and the Mt Billiemi–Mt della Fiera ridge. This hydrostructure presents a water flow mainly towards SE and W-NW. The Mirto spring is sited in the Mt Mirto hydrostructure.
- Belmonte–Pizzo Mirabella hydrostructure (Belmonte–Pizzo Mirabella water body): the water flow (Fig. 6.3) is oriented both NE and NW (in the direction of the spring S. Maria at Altofonte) and SE (Risalaimi spring group). The Santa Maria, Risalaimi, Dammusi and Frassino springs are placed in the Belmonte Mezzagno–Pizzo Mirabella hydrostructure (Fig. 6.4).

Mt Gradara Hydrogeological Unit

- *Mt Gradara hydrostructure (Mt Gradara water body): the Mt Gradara hydrostructure continues below the Mt Saraceno–Mt Cuccio Unit and presents a predominantly S to N groundwater flow, with a component directed towards N-NE as it approaches the Mt Saraceno relief. The Soprasansotta, Locogrande 1 and Giardinello springs are situated in the Mt Gradara hydrostructure.*

Mt Saraceno-Mt Cuccio Hydrogeological Unit

- *Mt Saraceno hydrostructure (Mt Saraceno water body)* constitutes a totally isolated structure surrounded by the Numidian Flysch. It presents divergent underground water flow towards N and it is responsible for the feeding of the springs located on the northern side of the relief.
- *Mt Cuccio–Mt Gibilmesi hydrostructure (Mt Cuccio–Mt Gibilmesi water body):* this hydrostructure as a whole represents a syncline buffered by the impermeable clays of the Numidian Flysch to the E (Plain of Palermo) and SE (Oreto Basin), which represent the groundwater discharge zones of the hydrostructure that occur in the springs of S. Martino delle Scale, Monreale, Baida-Boccadifalco and the Gabriele (Plain of Palermo). The *Gabriele* spring is in the Mt Cuccio–Mt Gibilmesi hydrostructure (Fig. 6.4 n. 2). Figure 6.4 shows the distribution of the springs in the various water bodies.

6.3 Hydrogeochemical Characterisation

In the Palermo Mountains area, 11 springs representative of 7 significant water bodies were sampled (Table 6.2, Fig. 6.5). In each sample, T, EC, and pH were determined in situ using portable instruments; chemical analyses of major, minor, and trace elements were performed in the laboratory. Temperature ranged between 13 and 18°C, conductivity between 350 and 550 μS/cm (at 20°C), and pH between 7.2 and 7.7 (Table 6.2).

In the Langelier-Ludwig classification diagram (Fig. 6.5), the waters of the sampled springs fall into the bicarbonate-alkaline-earthy quadrant. Also considering literature data, a clear trend toward the chlorinated-sulfate-alkaline water field is observed, moving from the chlorinated-sulfate-alkaline-earthy field.

The composition of the waters is perfectly compatible with the carbonate rocks massively present in the area. The trend towards the representative seawater composition could be related to the effects of marine intrusion (see Fig. 6.3).

Observation of the ternary diagrams (Fig. 6.6) shows a higher Ca content in the groundwater circulating in the hydrostructure north of the Mt Cuccio overflow, characterized by more calcareous rocks, and higher concentrations of Mg in the waters of the southern sector where dolomitic rocks are found (Mt Cuccio and Belmonte Mezzagno). The trend towards the vertexes of Na and Cl is attributable to both seawater ingression and ion exchange phenomena related to interaction with clays (Alaimo et al. 1984).

Previous studies (Cusimano et al. 1996; Calvi et al. 1998) report on the presence in the study area of a multi-stratum aquifer consisting of a surface aquifer, set in calcarenites with sandy and/or sandy-silt intercalations, and a deep aquifer set in Mesozoic limestones and dolomites. The deep aquifer in some sectors is confined by clayey levels of varying thickness, attributable to the Numidian Flysch or to Pleistocene sandy clayey-siltstone successions.

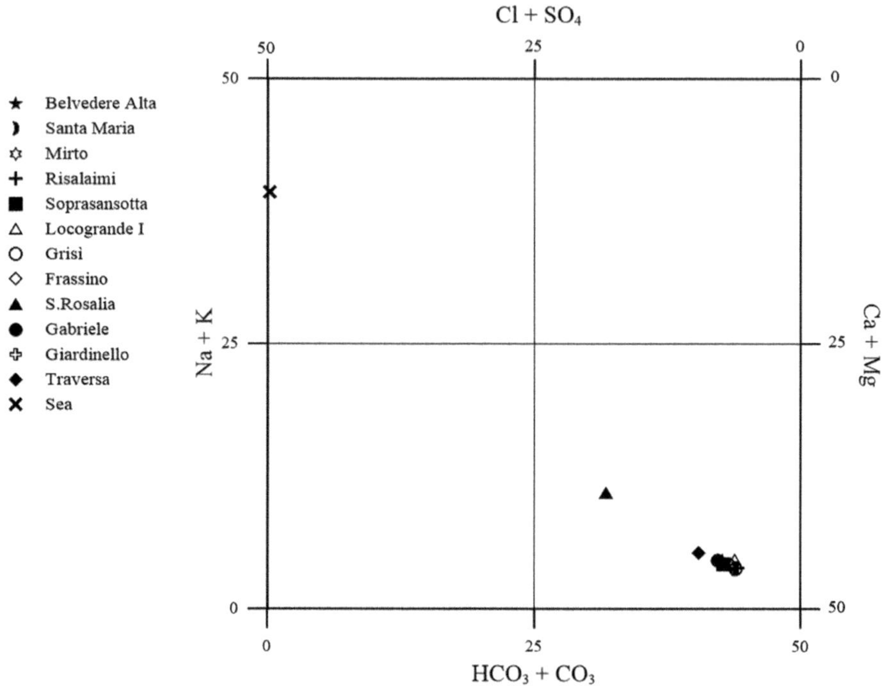

Fig. 6.5 Langelier–Ludwig classification diagram. Springs fall in the bicarbonate-alkaline-earthy, chlorine-sulphate- alkaline-earthy and chlorine-sulphate- alkaline quadrants. Literature data (INGV 2005) are shown for comparison; S: spring, W: well

6.4 Isotopic Characteristics of the Catchment Area

The drinking waters sampled in hydrological basin align on a line very close to the MMWL (Fig. 6.7). The LMWL for Palermo Mts basin, determined by Liotta et al. (2013) as sector 1 (see Cap. 1, Table 1.6) is $\delta^2H = 6.31\delta^{18}O + 5$. The stable isotope composition of groundwater ranges from −8.11 to −7.3 δ‰ for $\delta^{18}O$ and from −45 to −40 δ‰ for δ^2H (this study and INGV 2005). Collected and literature water data fall close to the LMWL, confirming a predominant meteoric origin of these waters.

According to the isotopic gradient of 0.15 δ‰/100 m (Table 1.7 relative to sector 1), the altitude of supply is higher than 700 m above sea level compatible with the average altitude of the basin (Fig. 6.8).

Table 6.2 Chemical and isotopic composition of the sampled points. Analytical data are expressed as meq/L, EC in uS/cm and δ^2H and $\delta^{18}O$ in ‰ vs SMOW

	T	EC	pH	Eh	Na	K	Ca	Mg	Cl	NO_3	SO_4	HCO_3	δ^2H	$\delta^{18}O$	WB
Santa Maria	14.5	455	7.37	251	0.4	0.017	3.22	1.89	0.44	0.041	0.20	4.80	−43	−7.89	BPM
Risalaimi	17.0	538	7.56	327	0.49	0.025	3.55	2.69	0.52	0.065	0.23	5.96	−42	−7.26	BPM
Grisì or Dammusi	15.2	441	7.72	242	0.39	0.014	3.19	1.88	0.44	0.019	0.21	4.80	−43	−7.89	BPM
Frassino	14.3	448	7.55	260	0.44	0.009	2.71	2.36	0.46		0.26	4.85	−43	−8.11	BPM
S. Rosalia	17.8	531	7.59	210	1.24	0.010	4.05	0.43	1.62	0.142	0.38	3.70	−34	−6.38	MC
Gabriele	15.8	491	7.29	298	0.54	0.020	2.97	2.65	0.57	0.133	0.25	5.15	−43	−7.7	MCMG
Soprasansotta	13.6	431	7.61	295	0.43	0.010	3.22	1.63	0.53	0.043	0.23	4.65	−42	−7.69	MG
Locogrande I	14.8	520	7.48	372	0.57	0.017	2.94	2.87	0.55	0.023	0.20	5.45	−45	−7.57	MG
Mirto	14.0	433	7.6	267	0.43	0.010	2.51	2.42	0.47	0.043	0.18	4.61	−45	−7.77	MM
Belvedere Alta	15.1	457	7.45	270	0.49	0.010	2.76	2.29	0.54	0.08	0.20	4.75	−40	−7.37	MP
Giardinello	14.6	530	7.26	288	0.53	0.030	3.34	2.98	0.54	0.127	0.33	5.79	−40	−8.06	MG
Traversa	13.5	370	7.59	361	0.41	0.010	3.44	0.22	0.4	0.139	0.24	3.31	−40	−7.85	MK

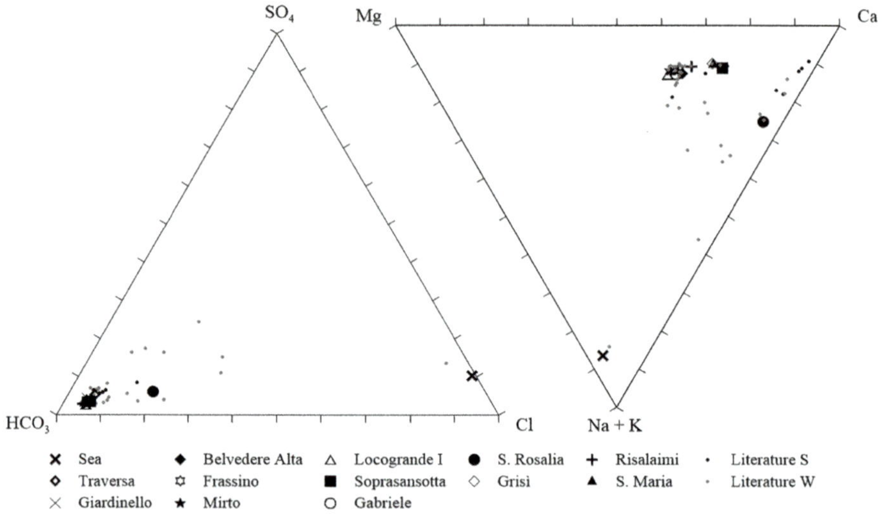

Fig. 6.6 Anions (left) and cations (right) ternary diagrams for Palermo Mts catchment area. Literature data are shown (INGV 2005). S: Spring, W: Well

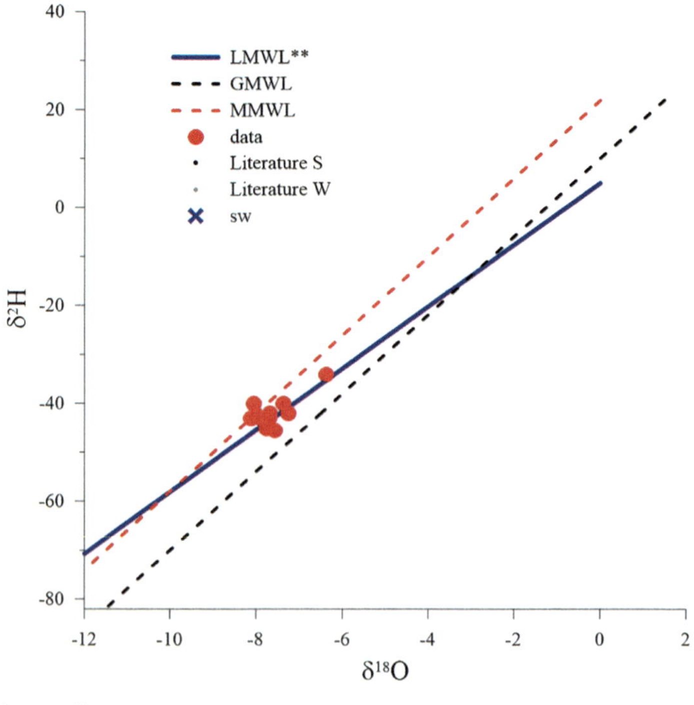

Fig. 6.7 $\delta^2H - \delta^{18}O$ (in ‰ versus SMOW) diagram of spring waters. The LMWL** is from Liotta et al. (2013). The GMWL and MMWL are shown for comparison. Literature data from INGV (2005). S: spring, W: well

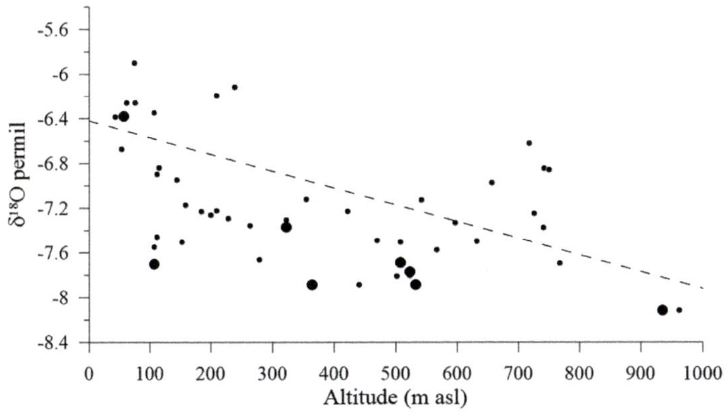

Fig. 6.8 Relationship between $\delta^{18}O$ and altitude

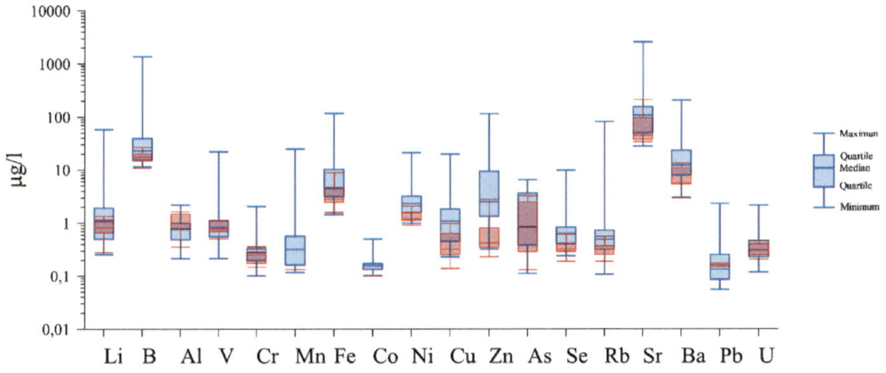

Fig. 6.9 Box-whisker plot of trace elements at Palermo Mts basin. Red box spring data (this work). Light blue box data from INGV (2005)

6.5 Trace Elements

Trace element concentrations in water springs (literature; INGV 2005) are rather wide but always below the legal limits (Table 1.1) for drinking waters. On the contrary, in sampled springs the range of variation is narrow (Fig. 6.9; Tables 6.2, Table 6.3).

Sr, Ba and B are the most concentrated elements, due to the carbonate nature of the rocks.

Table 6.3 Trace elements in sampled waters. Concentrations are in ug/L

	Li	B	Al	V	Cr	Mn	Fe	Co	Ni	Cu	Zn	As	Se	Rb	Sr	Ba	Pb	U	WB
S.Maria	0.89	14.9	1.47	0.51	0.37	bdl	4.78	bdl	1.35	0.14	0.81	3.33	0.33	0.376	94.8	5.6	bdl	0.313	BPM
Risalaimi	1.18	20.4	1.5	0.53	0.18	bdl	2.91	bdl	2.1	0.99	2.84	0.29	0.62	0.565	212	10.8	0.152	0.467	BPM
Dammusi 1	0.87	10.9	0.55	1.15	0.29	0.14	2.7	bdl	1.2	0.26	0.35	0.39	0.32	0.254	48.0	13.3	bdl	0.407	BPM
Frassino	1.37	19.7	1.66	0.83	0.15	bdl	2.19	bdl	1.25	0.64	0.76	1.35	0.19	0.19	43.0	3.0	bdl	0.391	BPM
S. Rosalia	0.81	27.4	0.8	1.12	0.28	bdl	2.94	0.1	1.6	0.46	0.23	2.55	0.3	0.337	96.9	8.4	bdl	0.251	MC
Gabriele	0.28	18.5	0.96	0.72	0.27	bdl	8.99	bdl	1.14	0.27	0.43	0.13	0.41	0.341	37.6	5.6	0.177	0.207	MCMG
Mirto	0.67	11.3	1.07	1.03	0.16	bdl	2.46	bdl	1.11	0.21	0.38	0.26	0.29	0.224	34.1	5.4	bdl	0.308	MM
Belvedere	0.41	17.1	0.36	0.7	0.25	bdl	1.6	bdl	0.92	0.38	bdl	1.57	0.36	0.305	33.7	4.3	bdl	0.216	MP

Bibliography and Recommended Further Reading

Abate B, Catalano R (1974) The margin of the Triassic Panormid Carbonate Platform in the Palermo Mountains. Boll Soc Nat Napoli 83:1–16

Abate B, Catalano R, Renda P (1978) Geological scheme of the Palermo Mountains. Boll Soc Geol It 97:807–819

Abate B, Catalano R, D'Argenio B, Di Stefano E, Di Stefano P, Lo Cicero G, Montanari L, Pecoraro C, Renda P (1982) Evolution of hinge zones between carbonate platforms and basins in western Sicily. In: Catalano R, D'Argenio B (eds), Guide to the Geology of Western Sicily, Regional Geological Guides. Mem Soc Geol It, Suppl. A., vol 24, pp 53–76, Palermo

Abate B, Incandela A, Nigro F, Renda P (1998) Plio-Pleistocene strike-slip tectonics in the Trapani Mts. (NW Sicily). Boll Soc Geol It 117

Agate M, Catalano R, Infuso S, Lucido M, Mirabile L, Sulli A (1993) Structural evolution of the Northern Sicily Continental Margin during Plio-Pleistocene. In: Max MD, Colantoni P (eds) Geological development of the Sicilian-Tunisian platform. Proceedings of the international scientific meeting held at the University of Urbino, vol 58. UNESCO Reports in Marine Science, pp 25–30

Alaimo R, Ferla P, Hauser S (1984) Hydrogeochemistry of the waters of the Palermo Plain. La Palma Editions, pp 79, Pa

Ambrosetti P, Bosi C, Carraro F, Ciaranfi N, Panizza M, Papani G, Vezzani L, Zanferrari A (1983) Neotectonic map of Italy (F. 5-6). C.N.R. P.F.G., neotectonic, L.A.C., Firenze

Aruta L, Buccheri G (1971) The preevaporitic Miocene in carbonate-detritic facies of the surroundings of Baucina, Ventimiglia di Sicilia, Calatafimi (Sicily). Riv Min Sic 130–132:188–194

Bigi G, Cosentino D, Parotto M, Sartori R & Scandone P (1991) Structural Model of Italy. (1/500.000). C.N.R. Geodynamics Finalized Project, Three-Dimensional Structural Model., Firenze

Broquet P, Duee G, Mascle G, Truillet R (1984) Recent alpine structural evolution in Sicily and its geodynamic significance. Rev Géol Dyn Et Géogr Phys 25(2):75–85

Caflisch L (1966) The Geology of the Palermo Mountains. Riv Ital Paleont Strat 12:108 pp

Calvi F, Contino A, Cusimano G, Di Cara A, Frias Forcada A, Hauser S, Pellerito S (1998) New contributions to the knowledge of the Hydrogeology of the Palermo Plain. Proceedings of the 79th National Congress of the Italian Geological Society., vol A. Palermo 21–23 September 1998, pp 212–215

Calvi F, Contino A, Cusimano G, Di Cara A, Frias Forcada A, Hauser S, Pellerito S (2000) Hydrostructures related to the Piana di Palermo aquifers and their hydrogeochimical characteristics. In: Catalano R, Lo Cicero G (eds) Sicily, a natural laboratory in the Mediterranean area: structures, seas, resources and hazards. Mem Soc Geol It 55:473–481

Calvi F, Contino A, Cusimano G, Di Cara A, Di Maggio C, Frias Forcada A, Hauser S, Pellerito S (2001) Impact of the Palermo Municipality solid waste disposal site on the surrouding aquifers an hydrogeological and hydrogeochemical approach. Italo-Russian institute of education and geological research. In: Proceeding of the "2nd symposium protection of groundwater from pollution and seawater intrusion", Bari, pp 225–239

Catalano R, D'Argenio B (1978) An essay of palinspastic restoration across the Western Sicily. Geol Romana 17:145–159

Catalano R, D'Argenio B (1982) Guide to the Geology of Western Sicily. Soc Geol It, Guide Geol Reg pp 155

Catalano R, Di Maggio C (1996) Tectonic overlap of the Imerian Units on the Panormids in the Palermo Mountains (Sicily). Natur Sic IV, XX(3–4):147–166

Catalano R, Lo Cicero G (eds) (1998) Sicily, a natural laboratory in the Mediterranean. Strutture, mari, risorse e rischi. 79th Congress Italian Geological Society. Palermo 18–20 september 1998, Hikes guide, vol 1, Western Sicily, pp 185

Catalano R, Maniaci G, Renda P, Urso G (1973) An example of sedimentary evolution in basin facies in the Palermo Mountains. The mesozoic-tertiary succession of Calarossa.(Terrasini). Geol Romana 12:151–175
Catalano R, D'Argenio B, Lo Cicero G (1974) The Triassic cyclotomes of Cape Rama (Palermo Mountains.). Geol Rom 13:125–145
Catalano R, Abate B, Renda P (1978) Geological map of the Palermo Mountains. Scale 1:50.000, Palermo
Catalano R, Franchino A, Merlini S, Sulli A, Agate M, Basilone L (1998) Materials for Understanding the Deep Layout of West-Central Sicily. In: Catalano R, Lo Cicero G (eds) "La Sicilia Occidentale", field trip guide of the 79° Congr. Naz. S. G. I., Mondello (Palermo), pp 175–185
Catalano R, D'Argenio B, Montanari L, Renda P, Abate B, Monteleone S, Macaluso T, Pipitone G, Di Stefano E, Lo Cicero G, Di Stefano P, Agnesi V (1979) Contribution to the knowledge of the structure of Western Sicily: The Palermo-Sciacca profile. Boll Soc Geol It 19:485–493
Catalano R, Franchino A, Merlini S, Sulli A (2000) Central western Sicily structural setting interpreted from seismic reflection profiles. Mem Soc Geol It 55:5–16
Contino A, Cusimano G, Frias Forcada A (1998) Hydrogeological Model of the Palermo Mountains. Proceedings of the 79th National Congress of the Italian Geological Society., Palermo 21–23 September 1998, vol A, 334–336
Contino A, Cusimano G & Frias Forcada A (1999) New contributions to the knowledge of the hydrostructural structure of the Palermo Mountains, "Geologists of Sicily" n 2 April-June 1999.
Contino A, Cusimano G, Di Cara A, Di Maggio C, Frias Forcada A, Hauser S (2001) Pollution of aquifers in the Monti di Palermo area: the role of geologic structure, kastification and urban impact. In: Mudry J, Zwaler F (eds) Proceeding of the 7th conference on limestone hydrology and fissured media, Besançon, France. Pubbl. n. 2308 G. N. D. C.I.–C. N. R
Contino A, Giammarinaro MS, Varsalona S (2003a) Southeastern sector of the city of Palermo: stratigraphic-geotechnical study of the portion not affected by alluvial deposits of the Oreto. "Geologists of Sicily.", anno XI, n 3, July-September 2003, 9–24
Contino A, Cusimano G, Di Cara A, Di Maggio C, Frias Forcada A, Hauser S (2003b) The role of karst in assessing the potential vulnerability to pollution of carbonate aquifers in the Palermo Mountains. Proceedings of the Conference "Karstification in the Mediterranean Area.", Castro Marina (Lecce), 14–15–16 september 2001, Department of Biological and Environmental Sciences and Technologies (Di.S.Te.B.A.), University of Lecce, vol 26 supplement of "Thalassia Salentina.", pp 319–327
Cusimano G, Frias Forcada A, Hauser S (1996) Salinization processes characterizing the aquifers of the Palermo Plain. Underground Waters, a. XIII, Geograph, Segrate (MI)
D'Angelo U, Ruggeri G, Unti M, Vernuccio S (1980) Neotectonics of Sheets 248 (Trapani), 249 (Palermo), 250 (Bagheria), 258 (Alcamo), 259 (Termini Imerese) e 266 (Sciacca). In: Contributions to the realization of the Neotectonic Map of Italy, pubbl. n 356, CNR-P. F. Geodinamica, 203–236, Roma
Di Maggio C (1997) Morphostructural arrangement and geomorphological evolution of a sector of the Palermo Mountains. Doctoral dissertation, University of Naples. "Federico II", pp 175
Di Maggio C (2000) Morphostructural aspects of the central northern sector of Palermo Mountains (Sicily). Mem Soc Geol It 55:353–361
Di Stefano P, Senowbari-Daryan B (1985) Upper Triassic Dasycladaces (Green Algae) from the Palermo Mountains (Sicily, italy). Geol Rom 24:189–220
Fabiani R, Trevisan L (1940) Evidence for the existence of a tectonic overlay layer style in the Palermo Mountains. Atti R Acc d' Italia, Mem Cl Sc Fis Mat e Nat 11:435–448
Ferla P, Bommarito S (1988) Middle Jurassic lateritic bauxites in the limestones of the Panormid carbonate platform of Monte Gallo (Palermo). Boll Soc Geol It 107:579–591
Ghisetti F, Vezzani L (1982) The role of the M. Kumeta-Alcantara shear zone in the structural evolution of the Calabrian Arc: implications and problems. In: Catalano R, D'Argenio B (Ed.), "Guide to the geology of Western Sicily", Mem Soc Geol It, suppl. A, 2, 119–123

Ghisetti F, Vezzani L (1984) Thin-skinned deformations of the Western Sicily Thrust Belt and relationships with crustal shortening: mesostructural data on the Mt. Kumeta-Alcantara fault zone and related structures. Boll Soc Geol It 103:129–157
Giunta G (1991) Elements for a kinematic model of Sicilian maghrebids. Mem Soc Geol It 47:297–311
Giunta G, Liguori V (1973) Paleotectonic evolution of northwestern Sicily.. Boll Soc Geol It 92:903–924
Giunta G, Nigro F, Renda P (2000) Extensional tectonics during Maghrebides chain building since late Miocene: examples from Northern Sicily. Ann Soc Geol Pol 70:81–98
Giunta G, Nigro F, Renda P, Giorgianni A (1998) A neotectonic model of the Tyrrhenian margin of the Sicilian Maghrebids (exten. abs). 79th Congress Italian Geological Society., Palermo (Italy) 21–23 September 1998, 498–501
Giunta G, Nigro F, Renda P, Giorgianni A (1999) Seismotectonic features in the NW Sicilian Maghrebides. Studi Geologici Camerti
Grasso M, Reuther CD (1988) The western margin of the Hyblean Plateau: a neotettonic transform system on the SE Sicilian foreland. Ann Tectonicae 2(2):107–120
Istituto Nazionale di Geofisica e Vulcanologia (INGV), Sezione di Palermo (2005) First characterization and monitoring of groundwater aimed at the drafting of the "Water Protection Plan of the Sicily Region". On behalf of the Sicilian Region - Deputy Commissioner for the waste emergency and water protection
Kiratzi AA (1994) Active seismic deformation in the Italian Peninsula and Sicily. Ann Geof 37(1):2–4
Liotta M, Grassa F, D'Alessandro W, Favara R, Gagliano Candela E, Pisciotta A, Scaletta C (2013) Isotopic composition of precipitation and groundwater in Sicily, Italy. Appl Geoch 34:199–206
Mascle G (1970) Geological sketch of Western Sicily. In: Alvarez W, Gohrbandt KHA (eds) Geology and history of Sicily, pp 231–243, Petr Expl Soc Libya, Tripoli
Montanari L, Rizzotto M (2000) Geology of Aspra. The Sicilian Naturalist 24:315–328
Montone P, Amato A, Frepoli A, Mariucci MT, Cesaro M (1997) Crustal stress regime in Italy. Ann Geof 40(3):741–757
Nigro F, Renda P (1999a) The North-Central Sicily belt: structural setting and geological evolution. Ann Soc Geol Poloniae
Nigro F, Renda P (1999b) Evidence of recent transcurrent tectonics in the Palermo Mountains (Sicily NW). The Quaternary 11(2):1–4
Nigro F, Renda P (1999c) Geological evolution and structural arrangement of north-central Sicily. Boll Soc Geol It 118:375–388
Nigro F, Renda P (1999d) Plio-Pleistocene Wrench Tectonics in the Western Sicily Chain. Ann Soc Geol Poloniae 69:99–112
Nigro F, Renda P (2000) A model of tectono-sedimentary evolution of the Sicilian Neogene foredeep. Boll Soc Geol It 119:667–686
Nigro F, Renda P (2004) The contribution of the pre-existing structures in the mountain belt evolution: the example of the negative inversion in Northern Sicily. Boll Soc Geol It 123:175–187
Nigro F, Renda P, Arisco G (2000) Recent tectonics in northwestern Sicily and the Egadi Islands. Boll Soc Geol It 119:307–319
Oldow JS, Channel JET, Catalano R, D'Argenio B (1990) Contemporaneous thrusting and large-scale rotations in the Western Sicilian fold and thrust belt. Tectonics 9:661–681
Patacca E, Sartori R, Scandone P (1992) Tyrrhenian Basin and Apenninic Arcs: kinematic relations since late Tortonian times. Mem Soc Geol It 45:425–451
Rehault JP, Moussat E, Fabbri A (1987) Structural evolution of the Tyrrhenian back-arc basin. Mar Geol 74:123–150
Ruggieri G, Sprovieri R (1970) Marls of S. Cipirello. Illustrative Studies of the Geological Map of Italy. Geological formations., Fasc. 4, 12
Ruggieri G, Milone G (1972) Tyrrhenian sign at 50 meters elevation near Palermo. Boll Soc Geol It 91:647–654

Scandone P, Radoicic R, Giunta G, Liguori V (1972) On the significance of Fanusi dolomites and ellipsatcinite limestones in northern Sicily. Riv Min Sic 23:51–61

Schutte KG (1978) Crustal structure of southern Italy. In: Closs H, Roeder D, Schmidt K (eds) Alps, Apennines, Hellenides. Stuttgart, pp 315–321

Torelli L, Zitellini N, Argnani A, Brancolini G, De Cillia C, Peis D, Tricart P (1991) Crustal geologic section from the Pelagian foreland to the Tyrrhenian backarc basin (central Mediterranean). Mem Soc Geol It 47:385–399

Wezel FC (1985) Structural features and basin tectonics of the Tyrrhenian Sea. In: Stanley DJ, Wezel FC (eds) Geological evolution of the Mediterranean Basin. Springer-Verlag, pp 153–194

Wezel FC, Savelli D, Bellagamba M, Tramontana M, Bartole R (1981) Plio-quaternary depositional style of sedimentary basins along insular Tyrrhenian margins. In: Wezel FC (ed) Sedimentary basins of Mediterranean margins. C.N.R. Italian Project of Oceanography, Tecnoprint, pp 239–269

Peloritani Mountains Hydrogeological Basin

7

Sabina Morici, Giovannella Pecoraino,
Esterina Gagliano Candela, and Leonardo La Pica

7.1 Geological Framework

The Peloritani Mountains constitute the extreme southern portion of the Calabrian-Peloritanian Arc. They develop along an alignment NE–SW from Cape Peloro up to the basin of the Alcantara (Fig. 7.1). They are constituted by a group of reliefs of more than 1000 m in height (Mt Grande 1376 m a.s.l.) that degrade towards the Ionic and Tyrrhenian coasts until they reach an altitude of 400–500 m.

The following is a description of the stratigraphic-structural units that make up the Peloritani Mountains, from bottom to top (Fig. 7.2):

Peloritanian Units

- *Longi-Taormina* Unit: constituted by an epimetamorphic basement covered by a Meso-Cenozoic sedimentary succession. It outcrops in the frontal sectors of the Peloritanian element according to a NW–SE trend from S. Agata di Militello to Taormina, along the southern front of the Peloritanian Mountains, marking the so-called "Taormina line".
- *Fondachelli* Unit: formed by a pre-alpine epimetamorphic basement composed of phyllites and metarenites. It emerges from the Mirto-Caprileone area on the Tyrrhenian coast, to Mt Galfa-Mt Veneretta near Taormina, on the Ionian coast.

S. Morici · G. Pecoraino (✉) · E. Gagliano Candela (Deceased) · L. La Pica
Istituto Nazionale di Geofisica e Vulcanologia, Sezione di Palermo, Via Ugo La Malfa 153, 90146 Palermo, Italy
e-mail: giovannella.pecoraino@ingv.it

S. Morici
e-mail: sabina.morici@ingv.it

L. La Pica
e-mail: leonardo.lapica@ingv.it

E. Gagliano Candela et al. (eds.), *Drinking Water in Sicily*,
https://doi.org/10.1007/978-3-031-60473-7_7

Fig. 7.1 The Peloritani Mountains. Imagine from Google Earth Pro v 7.3.6.9345 (Dicember 29, 2022). Sicily, Italy

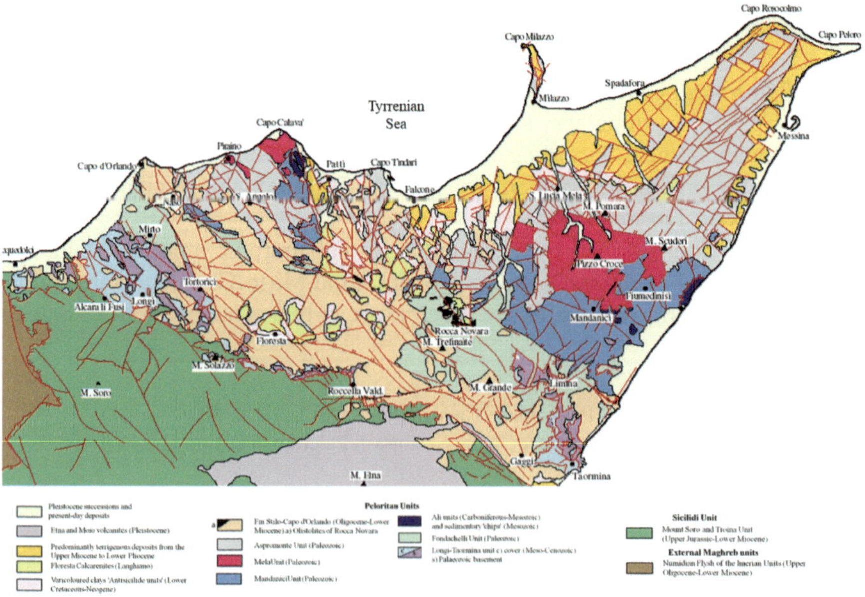

Fig. 7.2 Geological scheme of Peloritani Mountains (modified from Giunta et al. 1998)

- *Mandanici* Unit: represented by a pre-alpine basement composed of phyllites, quartzites, metabasites, marbles and porphyroids. The Unit emerges from the Fiumara of Mandanici on the Ionian side, up to the Fiumara of Naso, on the Tyrrhenian side.
- *Mela Unit: formed by a pre-alpine polymethamorphic basement consisting of paragneiss and micascists with intercalations of metabasites and marbles. The Mela Unit outcrops along a band with E–W direction, from Giampilieri Marina on the Ionian coast to Cape Calavà in the Tyrrhenian one.*
- *Aspromonte* Unit: formed by a metamorphic basement of medium–high grade, consisting of micascists and biotite paragneiss, in "Amphibolite" facies. The Unit occupies the entire northeastern portion of the Peloritani Mountains.

The Peloritanian units are separated by first-order tectonic contacts, along which, especially between the Fondachelli and Mandanici Units, and between these and the Aspromonte Unit, there are limbs of Meso-Cenozoic sedimentary successions ("tectonic shavings"). Of these limbs, the most extensive one emerges at Capo Alì.

Late-orogenes, pre-collisional covers of the Oligocene-low Miocene

- In discordance with the tectonic units described, the **Capo d'Orlando Flysch**, consisting of fine yellowish-grey micaceous sandstones and marly clays and grey or brown marls, with conglomerates with predominantly granitic-gneissic elements, emerges extensively. It occurs in the south-western sector of the Peloritani Mountains.

Antisicilide Unit

- the sedimentation of the Capo d'Orlando Flysch was interrupted during the Langhiano period following the tectonic covering by a blanket of ***Antisicilide Variegated Clays***. They have the appearance of polychrome clays with a flaky texture, often in a chaotic layout, interspersed with bituminous argilloschistes alternating with variegated radiolarites, and clayey-arenaceous layers incorporating yellowish quartzarenites. The quartzarenite inclusions vary in size from a few centimetres to several tens of metres.

Late-orogenes, pre-collisional covers of the Low-medium Miocene

- In discordance with the Antisicilide Units, and the underlying soils, lies the Miocene Unit of ***Floresta Calcarenites***. These consist of arkosic sandstones, with abundant carbonate cement bioclasts. The sandstones consist of sharp-edged quartz granules, mica, feldspar and lithics of metamorphic derivation. Their age is Langhiano. The calcarenitic facies are stratigraphically covered by a clayey-sandy horizon 50–70 m thick of the Upper Langhiano.

Late-orogenes, pre-collisional covers

- ***Tortonian-Messinian*** deposits are predominantly present in the northern part of the Peloritani Mountains and rest discordantly on all the oldest soils. The succession consists of sands, silts and conglomerates with various degrees of cementation. The lower portion of the succession is markedly distinguished by conglomeratic horizons alternating with sandy levels of metric thickness, while towards the stratigraphic top the pelitic-sandy facies is increasingly dominant.
- In the Capo Milazzo area and in the Nebrodi Mountains between Santo Stefano di Camastra and Caronia, discontinuous patches of *Porites* and Melobesia algae bioconstructions outcrop, discordant either on the crystalline basement or on the foredeep successions of the Numidian basin.
- Above this, thin, discontinuous limbs of evaporites are locally present. The succession begins with a thin and very discontinuous tripolaceous interval, followed by limestones, gypsums, chalky clays and brecciated clays, with vertical associations of variable facies.
- The latter are succeeded by marls and marly limestones, sometimes sandy (Trubi *Auct.*), marly clays, cross-stratified or plan-parallel calcarenites and ochre-yellow, Plio-Pleistocene organogenic sands.
- Marine terraces are quite widespread and highlight the intense uplifts to which the area was subjected in recent geological times and which may still be in progress.
- Along the Ionian and Tyrrhenian coasts there are alluvial deposits, both current and recent, and littoral deposits, mainly sandy, of fluvial-marine origin.

7.2 Hydrogeology

The complex stratigraphic and structural configuration of the Peloritani Mountains conditions the hydrogeology of the area, resulting in a rather inhomogeneous distribution of groundwater resources.

The water bodies of the Peloritani Mountains have different hydraulic characteristics and potential, in relation to the degree and type of permeability of the rocky media and their geometric characteristics (hydrostructures). The permeability of the outcropping soils varies in relation to both the primary (porosity) and secondary (cracking) characteristics, the latter depending on the degree of tectonisation of the rocky bodies.

A distinction can be made between terrains with different permeability:

- ***very high permeability due to primary porosity***: alluvial deposits at the bottom of the valleys of streams, rivers and coastal plains, dejection conoids and particularly thick detrital prisms;
- ***medium to high permeability due to primary porosity***: the sands and gravels of Messina, calcarenites and Pleistocene sands;

- ***high permeability due to cracking***: to this category belong the carbonate successions of the Longi-Taormina Unit, the conglomeratic facies of the Capo d'Orlando Flysch and the crystalline bodies of high metamorphic grade. Where they are permeated by fault networks and neotectonic fractures, permeability can become very high;
- ***medium–high permeability due to porosity and/or cracking***: the proximal arenaceous facies of the Capo d'Orlando Flysch, the peloritanian tardorogenic successions, the evaporitic deposits;
- ***medium permeability due to cracking and/or porosity***: arenaceous-clayey alternation of the upper Miocene, the Trubi and the medium grade metamorphites;
- ***medium–low permeability due to porosity and/or cracking***: terraced fluvio-marine deposits, distal clay-sandstone alternations of the Capo d'Orlando Flysch and low grade metamorphites;
- ***very low permeability***: blue marly clays, diatomites and marls, variegated clays, the anti-sicilide clay successions and the low-grade metamorphites.

Based on the permeability, location and geometry of the lithostratigraphic units, aquifers of varying potential are identified in the area.

The main aquifers are contained in the alluvial deposits at the bottom of the valleys of the torrents, in the form of independent bodies of water, which unify at the extensive coastal plain of Barcellona-Milazzo. The alluvial aquifers present at the bottom of the valleys that cut into the northern and eastern slopes of the Peloritani Mountains are generally similar to single-layer systems with a free water table, often hydraulically connected to watercourses. The feeding areas are represented by the catchment basins of the various water bodies with their mouths on the Tyrrhenian or Ionian coast. The groundwater recharge is essentially due to direct precipitation, snowfall at the highest altitudes, surface runoff along the riverbeds and the return of water infiltrated into the soils with discontinuous permeability and which comes to light in correspondence with the numerous springs. A further contribution is made at lower altitudes by the infiltration of water used for irrigation and wastewater from built-up areas. At the mouth of the watercourses in the coastal strips or in the Barcelona-Milazzo plain, the presence in the alluvial deposits of low-permeability intercalations, sometimes of significant areal extension, although not continuous, leads to conditions of semi-confinement of the water tables, configuring a compartmentalised type system.

In the case of the Barcelona-Milazzo plain there are also deep aquifers confined to calcarenitic or calcareous aquifers underlying pelitic sediments.

Aquifer levels of modest or negligible interest are contained in the fluvial alternations; they are characterised by limited extension and potential, since permeability is localised in the cracked arenaceous levels or banks, which alternate with practically impermeable clay layers, and water circulation is limited to the stone horizons. These horizons take on the role of natural reservoirs depending on the degree of cracking and the bedding conditions. Each of these horizons often behaves independently of similar sandstone bodies. However, in some cases, for tectonic reasons, interconnections between different horizons can also occur, increasing the volume of the reservoir and thus its storage capacity.

Similar hydrogeological significance can be found in the water bodies located in metamorphites, the existence of which is strictly dependent on the state of fracturing and surface alteration of these soils; they therefore constitute discontinuous aquifers in the form of levels of varying potential that feed springs generally of modest capacity (Ferrara 1999).

In calcareous aquifers and clayey sandstone alternations, water circulation is strongly conditioned by the degree of fracturing, although porosity also plays a significant role. The flow directions are regulated by the position of the lithological complexes resting on the metamorphites, with frequent immersion of these towards the north and north-east in monoclinal structures.

Other aquifers of non-negligible importance, even if their presence is limited to a few sectors of the territory, are those present in evaporitic soils and in sands and calcarenites. The bedrock characteristics imply the existence of local hydraulic connections between these and the alluvial aquifers above. They can also be both free and confined aquifers, depending on the presence or absence of impermeable soils with sufficient continuity at the top.

The aquifers present in the evaporitic deposits are mainly made up of carbonate sediments (evaporitic limestone) which are of a certain interest on the Tyrrhenian side of the Peloritanian chain in relation to their thickness and position, as well as the high permeability due to the porosity that characterises these deposits. The water circulation is rather active, giving rise in the hilly areas to free water tables of limited extension due to the discontinuity of the deposits caused by depositional and/or tectonic reasons. Their productivity is locally of some interest, also due to interconnections with the sandy-calcarenitic aquifers above and below. However, from the point of view of the volume of water resources stored, these aquifers play a decidedly marginal role in the overall availability of the territory (Ferrara 1999).

The aquifers of sandy-calcarenitic deposits are well represented on both sides of the Peloritani Mountains, along the hilly belts bordering the crystalline rocks. These are complex aquifers due to the marked lithological variability that characterises them, resulting in the existence of discontinuous circulation within them. The structure is given by various levels of permeability, by porosity and/or by cracking, partially separated by semi-permeable levels of limited extension. Therefore, in these aquifers the vertical permeability is definitely lower than the horizontal permeability and the aquifers contained in them are partly free and partly confined, with discrete to mediocre potential. In the Peloritani Mountains, 19 water bodies have been recognised:

- North-Western Peloritans
- Western Peloritans
- Central Peloritans
- North-Eastern Peloritans
- Eastern Peloritans
- South-Eastern Peloritans

- Southern Peloritans
- Floresta
- Mirto-Tortorici
- Fondachelli-Pizzo Monaco
- Piana di Barcellona- Milazzo
- Alcantara
- S. Agata Di Militello—Capo d'Orlando
- Brolo
- Gioiosa Marea
- Timeto
- Roccalumera
- Messina-Capo Peloro
- Naso.

Table 7.1 shows a hydrogeological scheme of the water bodies with the hydrogeological complex they belong to, a summary description of the lithotypes and the type of aquifer.

In the Peloritani Mountains, there are numerous springs (Fig. 7.3) seeping out at different altitudes. These springs are connected, as already mentioned, to aquifer horizons of limited extension and modest thickness. Their waters, most of which are not captured because of the low flow rates and the great dispersion of the emergence points, flow down the slopes and feed the alluvial deposits at the bottom of the valley.

The greatest number of springs fall on the outcrops of crystalline soils and those with the most significant flow are located on the high grade metamorphites (eyed gneisses, biotite paragneisses with lenses of aplites and pegmatites), which make up the summit parts of the Peloritan chain. The springs with a higher flow rate and perennial character are captured by simple and often rudimentary systems and the water is used for drinking and irrigation purposes. Several temporary springs are also affected by simple water derivation systems, which are used to irrigate small lots of cultivated land. These are generally springs with a defined permeability limit (Civita 1972), in the case of springs located at the contact between carbonate or arenaceous terms and pelitic terms of the flyscioid alternations.

Springs with an imposed permeability threshold are sometimes found at the contact between high grade metamorphites and semimetamorphites of different structural units. Springs with undefined permeability limits are frequently found at the contact between crystalline and metamorphic soils and detrital and alteration covers.

Springs originating from crystalline soils often have non-negligible initial flow rates, but tend to be exhausted in a relatively short time due to the small size of the reservoirs.

The springs connected to terrigenous deposits and to flyschioid alternations mainly have modest flow rates and little persistence, being fed by aquifer horizons consisting of single levels or arenaceous banks that are generally of little extension.

Table 7.1 Scheme of water bodies in the Peloritani Mountains

Water body	Hydrogeological complex	Mouton hydrogeological complex	Description	PTA hydrogeological complex	Type of aquifer
Brolo	VF valley floods	Complex of alluvial deposits of major rivers	Fluvial deposits	Flood complex	Free aquifer
Gioiosa Marea	VF valley floods	Complex of alluvial deposits of major rivers	The alluvium of the Zappardino stream consists of medium to high grade metamorphic elements drained by the reliefs within the catchment area. The grain size of the deposits is generally represented by gravels and pebbles, with more or less sandy and silty lenses present in the final section of the alluvial bed	Flood complex	Free aquifer
Messina-Capo Peloro	VF valley floods	Complex of alluvial deposits of major rivers	It consists of the Ghiaie di Messina, in places covered by recent alluvial deposits. On the whole, these are gravelly-sandy deposits	Flood complex	Free aquifer
Roccalumera	VF valley floods	Complex of alluvial deposits of major rivers	It constitutes a water body of good hydrogeological interest, consisting of alluvial, coastal and Ghiaie di Messina deposits outcropping in the sector between Alì Terme and S. Alessio Siculo	Flood complex	Free aquifer

(continued)

Table 7.1 (continued)

Water body	Hydrogeological complex	Mouton hydrogeological complex	Description	PTA hydrogeological complex	Type of aquifer
S. Agata-Capo D'Orlando	VF valley floods	Complex of alluvial deposits of major rivers	The coastal plain is characterised by Holocene alluvial deposits expressed predominantly by sandstone and/or microconglomeratic grain size fractions, interspersed with levels characterised by a greater presence of fine-grained portions	Flood complex	Free aquifer
Timeto	VF valley floods	Complex of alluvial deposits of major rivers	The alluvium of the Timeto stream consists of medium to high grade metamorphic elements and sandstones drained from the elevations within the catchment area	Flood complex	Free aquifer
Barcellona-Milazzo Plain	VF/DEB Plio-quaternary alluvial floods and debris formations	Complex of alluvial deposits of major watercourses/sands complex	Coastal deposits show a prevalence of coarse elements, represented by gravel and pebbles, and a lower content of medium to fine-grained fractions, in contrast to the alluvium of the backlands where the silty-clayey fraction component is abundant	Complex of recent alluvium and dune sands, and Sicilian sands and gravels	Free aquifer

(continued)

Table 7.1 (continued)

Water body	Hydrogeological complex	Mouton hydrogeological complex	Description	PTA hydrogeological complex	Type of aquifer
North-Western Peloritans	CA/LOC limestones local aquifers	Limestone complex/ flysch complex/ crystalline rock complex	Mesozoic carbonate successions, Oligo-Miocene foreland terrigenous deposits, Frazzanò and Stilo-Capo d'Orlando Formations	Limestone complex/ Granitic-metamorphic complex/Flyscioid complexes	Free and confined aquifer
Western Peloritans	CA/LOC limestones local aquifers	Limestone complex/ flysch complex/ crystalline rock complex	Mesozoic-tertiary carbonate-terrigenous successions; Palaeozoic crystalline bodies represented by chloritic and sericitic phyllites, graphitic places, metarenites, metacalcites and semiscales	Limestone/ Granitic-Metamorphic Complex	Free and confined aquifer
· South-Eastern Peloritans	CA/LOC Limestones Local aquifers	Limestone Complex/ Flysch Complex/ Crystalline Rock Complex	Carbonate successions of the Longi-Taormina Unit. The oldest soils are composed of chloritic and/or sericitic phyllites, in graphitic places, with lenses of macrocrystalline metacalcars, associated with sericitic semiscists	Limestone/ Granitic-Metamorphic Complex	Free and confined aquifer

(continued)

Table 7.1 (continued)

Water body	Hydrogeological complex	Mouton hydrogeological complex	Description	PTA hydrogeological complex	Type of aquifer
Naso	DET Plio-Quaternary detrital formations	Sand complex	The basal part of the Pleistocene successions consists of weakly stratified yellow sands with thin and rare calcarenitic intercalations containing macrofossils. Towards the top the predominantly sandy lithotypes change to alternating layers of yellowish calcarenites and ochre sands	Sandy-calcarenitic	Free aquifer
North-Eastern Peloritans	DET/LOC Plio-quaternary detrital formations local aquifer	Plio-quaternary detrital formations/flysch complex/crystalline rocks complex	Upper Miocene-Pleistocene succession of sands, conglomerates and pelitic levels, resting on a clay substratum of the A. V. Antisicilides, in turn covering the crystalline basement of the Aspromonte Unit	Granitic-metamorphic complex /Flyscioid complexes	Free and confined aquifer
Floresta	LOC local aquifer	Flysch complex	The aquifer is contained within the Langhiane successions of the Floresta Calcarenites, i.e. arkosic carbonate cement sandstones with glauconite and abundant organic remains	Clay-silica-marly flyscioid complexes of Permo-Triassic-Oligomiocene age	Free aquifer

(continued)

Table 7.1 (continued)

Water body	Hydrogeological complex	Mouton hydrogeological complex	Description	PTA hydrogeological complex	Type of aquifer
Central Peloritans	LOC Local Aquifer	Flysch complex/ Crystalline Rock Complex	Crystalline rocks of medium to high metamorphic grade and from the terrigenous successions of the Stilo-Capo d'Orlando Formation	Granitic-metamorphic complex /Flyscioid complexes	Free and confined aquifer
Southern Peloritans	LOC Local Aquifer	Flysch complex/ Crystalline Rock Complex	Conglomeratic-arenaceous successions, covering the crystalline rocks of the Fondachelli Unit	Granitic-metamorphic complex /Flyscioid complexes	Free aquifer
Eastern Peloritans	LOC local aquifer	Crystalline rock complex	The body of water consists mainly of the metamorphites of the Aspromonte Unit on which lie thin and discontinuous patches of turbidites of the Stilo-Capo d'Orlando Formation, A. V. Antisicilides and deposits of the Upper Miocene	Granitic-metamorphic complex	Free and semiconfined aquifer
Alcantara	LOC/VU/VF local aquifers volcanites valley floods	Flysch complex/ Complex of pyroclastites, lavas and volcanites in general/ complex of alluvial deposits of major watercourses	Metamorphic rocks of the Calabrian Peloritanian Arc—ordered in systems of strata of crystalline and metamorphic soils and sedimentary units given by flaky clays, overlapping with the Monte Soro Flysch and Numidian Flysch, and with Etna's volcanites	Clay-silica-marly flyscioid complexes of Permo-Triassic-oligomiocene age/volcanite complexes/recent deposits	Free aquifer

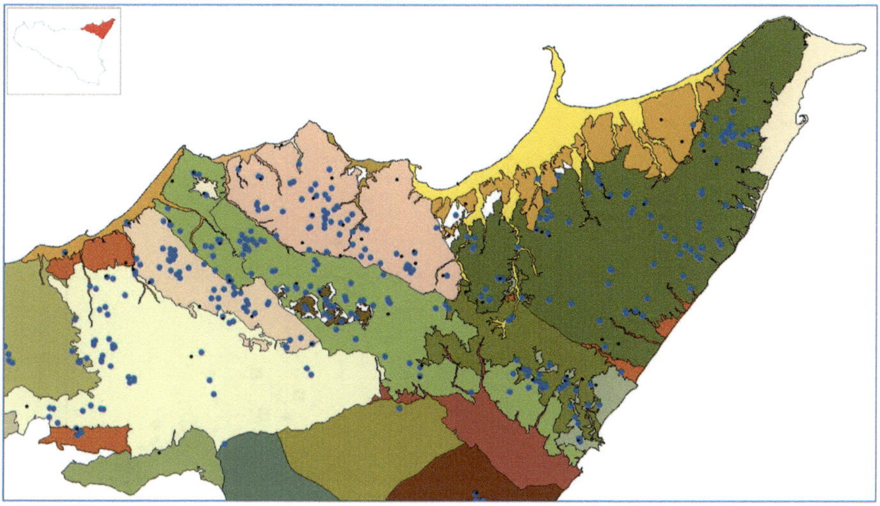

Fig. 7.3 Springs for drinking water use in the Peloritani Mountains (blue dots), black dots indicate springs used for characterization

7.3 Hydrogeochemical Characterisation

The factors regulating the content of dissolved species in the groundwater in Peloritani Mountains are essentially linked to the interaction of meteoric recharge waters with atmospheric particulates, rocks constituting the aquifers and, in areas close to the coastline, mixing with seawater.

In the north-eastern sector of Sicily, the most significant water bodies fall within the quartzarenitic portion of the flyschoid lithotypes (Capo D'Orlando Flysch), in the alluvial plains and metamorphites. The alluvial plains sometimes branch off inland in correspondence with the river courses, while the metamorphites differ significantly depending on the degree of metamorphism and their alteration. The lithological heterogeneity characterising the Peloritan aquifers provides very different geochemical characteristics, either between various water bodies or sometimes even within the same water body.

For the geochemical characterisation of the springs in this basin, 36 samples representative of 19 significant water bodies were considered. Literature data are included for comparison.

Water temperature ranges from 9.5 to 18.4 °C, the electric conductivity (EC at 20 °C) from 150 to 1217 μS/cm and the pH from 6.2 to 8.2.

In the Langelier-Ludwig classification diagram (Fig. 7.4), sample points plot in three quadrants reflecting that the water chemistry is diversified by different water–rock interaction processes and/or other contributions (such as a marine aerosol

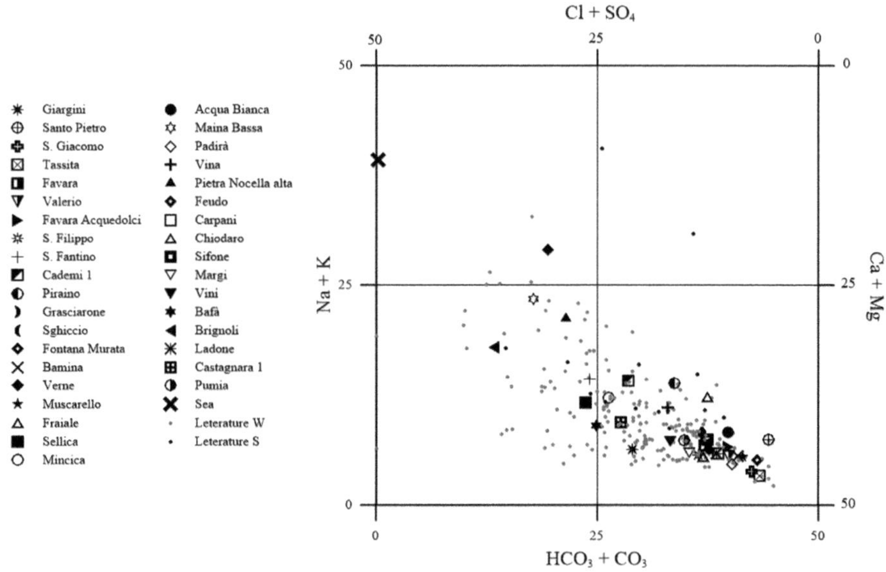

Fig. 7.4 Langelier—Ludwig classification diagram. Springs fall in the bicarbonate-alkaline-earthy, chlorine-sulphate- alkaline-earthy and chlorine-sulphate-alkaline quadrants. Literature data (INGV 2005) are shown for comparison; S: spring, W: well

input). Most of the samples fall in the bicarbonate-alkaline-earthy water quadrant, whereas some samples fall in the chlorinated-sulphated-alkaline-earthy water quadrant, and only one in the chlorinated(-sulphated) -alkaline water quadrant.

The most abundant components between anions and cations are the carbonate and calcium ions, respectively (Fig. 7.5). A shift towards alkaline ions and Mg (less extent), SO_4 and Cl (Fig. 7.6) is also recognised. The presence of crystalline limestone in the metamorphites and calcareous cement sandstone in the flyschoid successions and a light influence of marine aerosol justify these observations. The stationarity in chemical characteristic (INGV 2005) in time would indicate big, or underexploited water volume.

7.4 Isotopic Characteristics of the Catchment Area

The Peloritani hydrogeological basin is one of the rainiest areas in the entire region. The morphology determines, in fact, an orographic barrier to disturbances coming from the north-western quadrants, favouring the formation of both liquid and solid precipitation. Similarly to Mts Nebrodi (Chap. 5), the Peloritani hydrogeological basin is characterised by a dorsal ridge with elevation around 1000 m a.s.l. developing along a SW-NE trend. Also in this case, this involves the distinction of two zones with different climatic characteristics: one along the coastal lines, with thermal regime typical of Mediterranean areas, the other, innermost, typical of a

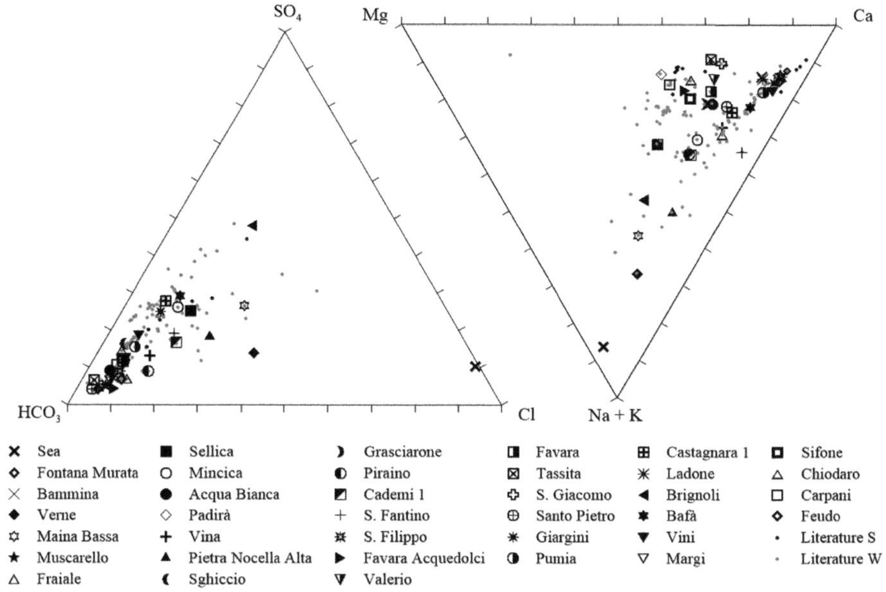

Fig. 7.5 Anions (left) and cations (right) ternary diagrams for Peloritani Mountains catchment area. Literature data are shown (INGV 2005). S: Spring, W: Well

mountain ridge, where the climate becomes more continental. These differences affect the isotopic composition of precipitation and therefore of groundwater.

The isotopic composition of groundwater varies from −8.8 to −5.9‰ for $\delta^{18}O$ and from −49 to −29‰ for $\delta^{2}H$ (Table 7.2) with the most negative values in high altitude areas. Sampled spring waters align very close to the MMWL line (Fig. 7.6) and on the LMWL for the Nebrodi and Peloritani basins (Liotta et al. 2013) as sector 3 (see Table 1.6). The LMWL is $\delta^{2}H = 6.14\ \delta^{18}O + 3.9$.

Collected and literature data fall close to the LMWL, confirming a meteoric origin for these waters. In particular, the marked difference between the isotopic compositions of sites located at higher elevation versus those closer to the coastline indicates direct recharge of low-elevation sites.

According to the isotopic gradient of 0.13δ‰/100m (Table 1.7 relative to sector 3, Nebrodi and Peloritani), the average altitude of supply is around 1200 m a.s.l., compatible with the average altitude of the basin (Fig. 7.7).

Waters located in the south-western sector of the basin show a more negative isotopic composition, compatible with feeding at higher elevation, than those located in the north-eastern sector. This behaviour is probably related to a contribution from the northern sector of the Nebrodi Mountains where elevations compatible with the observed values are present.

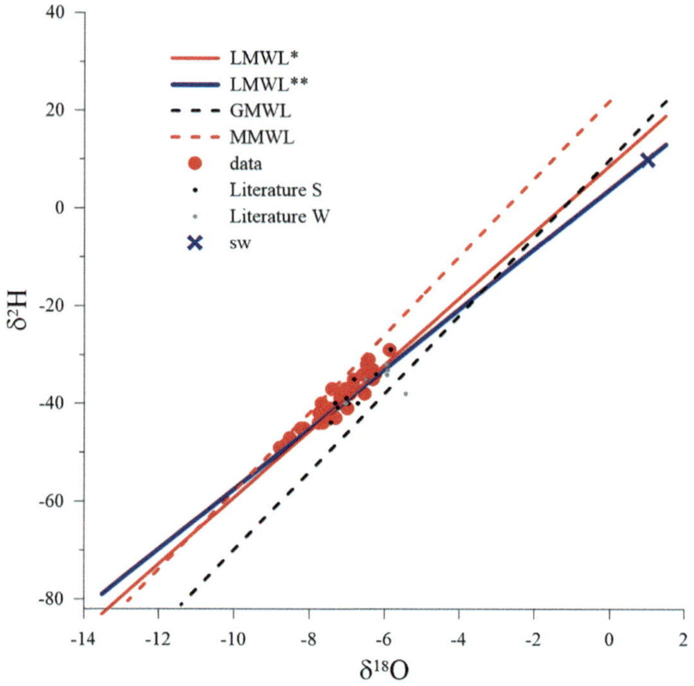

Fig. 7.6 δ^2H–δ^{18}O (in ‰ vs SMOW) diagram of spring waters. The LMWL* is from INGV (2005), LMWL** is from Liotta et al. (2013). The GMWL and MMWL are shown for comparison. Literature data from INGV (2005). S: spring, W: well

7.5 Trace Elements

Despite being a large-scale mining site in the past due to the presence of polymetallic mineralisations (Ferla and Meli 2007), in the SE area of Peloritani Mts (Mandanici Unit), the drinkwaters do not show anomalous, above-legal limits, concentrations of trace elements. The composition of trace elements in the waters under study varies in a range from 0.1 μg L^{-1} to 1 mg L^{-1}. Sr is by far the most concentrated element (Fig. 7.8).

Table 7.2 Chemical and isotopic compositions of the sampled points

	T	EC	pH	Eh	Na	K	Ca	Mg	Cl	NO_3	SO_4	HCO_3	δ^2H	$\delta^{18}O$	WB
Pietra Nocella Alta	13.7	153	6.87	206	0.49	0.024	0.45	0.25	0.35	0.117	0.21	0.51	−40	−7.67	FPM
Vina	11.9	262	7.53	195	0.61	0.046	1.73	0.59	0.56		0.44	1.95	−42	−7.57	MT
Padirà	11.3	427	7.76	162	0.47	0.028	2.36	2.48	0.53	0.051	0.46	4.3	−45	−8.15	MT
Maina Bassa	15.7	316	6.18	271	1.54	0.055	0.87	0.93	1.23	0.083	0.88	1.2	−39	−7.17	MT
Acqua Bianca	15.1	629	7.33	197	1.27	0.073	4.58	2.22	0.67	0.102	0.86	6.34	−41	−6.99	PC
Mincica	14.6	567	7.32	205	1.49	0.132	3.34	1.72	1.21	0.036	1.87	3.46	−37	−7.05	PC
Sellica	13.0	550	7.77	204	1.26	0.21	2.47	2.39	1.39	0.31	1.63	3.00	−39	−7.12	PC
Fraiale	17.8	288	7.73	271	0.77	0.025	1.86	0.59	0.51	0.031	0.26	2.39	−44	−7.64	PC
Muscarello	13.9	542	7.84	198	0.71	0.043	5.3	0.69	0.69	0.05	0.42	5.6	−42	−7.46	PC
Verne	12.2	150	6.83	222	0.6	0.025	0.23	0.22	0.5		0.14	0.41	−45	−8.22	PC
Bammina	15.4	509	7.30	189	0.65	0.055	4.66	1.02	0.63	0.117	0.41	5.2	−38	−6.53	PC
Fontana Murata	13.9	566	7.45	176	0.68	0.05	5.94	0.61	0.59	0.03	0.35	6.15	−41	−7.60	PC
Scriccio	16.2	644	7.37	197	1.25	0.065	3.55	3.22	0.6		1.59	6.05	−38	−6.92	PM
Grasciarone	11.2	340	7.83	184	0.61	0.044	2.18	1.19	0.5		0.56	3.00	−41	−7.44	PM
Piraino	12.8	202	7.31	203	0.45	0.023	0.79	0.44	0.36	0.02	0.17	1.13	−37	−7.40	PM
Cademi	11.8	261	7.65	199	0.78	0.09	1.36	0.72	0.72		0.53	1.66			PM
S. Fantino	18.4	1217	6.92	194	3.87	0.53	9.82	1.19	2.85	2.21	2.68	7.25	−35	−6.30	PNE

(continued)

Table 7.2 (continued)

	T	EC	pH	Eh	Na	K	Ca	Mg	Cl	NO_3	SO_4	HCO_3	δ^2H	$\delta^{18}O$	WB
S. Filippo	15.7	481	7.34	240	0.6	0.059	4.66	0.52	0.67	0.328	0.56	4.20	−33	−6.31	PNE
Favara (Acquedolci)	18.2	498	7.39	238	0.79	0.033	3.08	2.35	0.85	0.124	0.32	5.00	−38	−6.98	PNE
Valerio	14.8	322	7.85	228	0.4	0.029	2.26	1.2	0.41	0.09	0.28	3.1	−43	−7.30	PE
Favara	15.6	483	7.39	235	0.75	0.06	3.37	1.77	0.65	0.071	0.8	4.39	−42	−7.74	PE
Tassita	10.1	386	7.72	213	0.3	0.026	2.82	1.69	0.24	0.02	0.38	4.20	−49	−8.77	PE
S. Giacomo	9.5	339	7.98	194	0.29	0.028	2.56	1.31	0.36	0.021	0.24	3.51	−47	−8.52	PE
Santo Pietro	13.3	321	7.69	264	0.36	0.219	2.44	0.89	0.24		0.2	3.51	−44	−7.75	PE
Giargini	14.4	342	7.79	183	0.45	0.037	2.96	0.41	0.52	0.032	1.05	2.20	−38	−7.00	PE
Pumia	13.2	407	7.62	172	0.63	0.068	3.5	0.59	0.59		0.87	3.34			PE
Castagnara 1	16.1	296	7.41	479	0.58	0.041	2.05	0.64	0.43	0.022	1.00	1.80	−38	−7.16	PE
Ladone	10.8	255	8.20	177	0.33	0.022	2.17	0.45	0.31	0.128	0.25	2.30	−41	−7.5	PE
Brignoli	13.0	193	7.33	273	0.58	0.099	0.59	0.63	0.48	0.008	0.92	0.51	−35	−6.83	PE
Bafà	15.1	367	7.95	242	0.67	0.057	2.77	0.58	0.67	0.101	1.26	2.02	−34	−6.26	PE
Vena	10.9	278	7.95	165	0.4	0.052	2.39	0.29	0.35	0.037	0.67	2.10	−42	−7.57	MT
Margi	15.9	724	7.61	188	0.99	0.088	7.15	0.88	0.97	0.401	1.26	6.34	−31	−6.41	PSE
Sifone	16.7	494	7.69	309	0.84	0.052	3.05	2.13	0.63	0.04	0.88	4.61	−29	−5.86	PSE
Chiodaro	15.5	374	8.03	203	0.45	0.031	2.3	1.68	0.38	0.031	0.77	3.34	−34	−6.58	PSE
Carpani	14.9	323	8.04	190	0.43	0.021	1.79	1.68	0.38	0.008	0.49	3.00	−32	−6.47	PSE
Feudo	15.1	621	7.10	225	0.951	0.031	6.09	0.700411522633745	1.08	0.239	0.6	5.8	−36	−6.85	N

Analytical data are expressed as meq/L, EC in μS/cm and δ^2H and $\delta^{18}O$ in ‰ versus SMOW. WB is for water body. FPM is for Fondachelli-Pizzo Monaco, MT is Mirto-Tortorico, PC is for Central Peloritani, PM is for Southern Peloritani, PNE is for North-eastern Peloritani, PE is for Eastern Peloritani, PSE is for South-eastern Peloritani, N is for Naso, water bodies

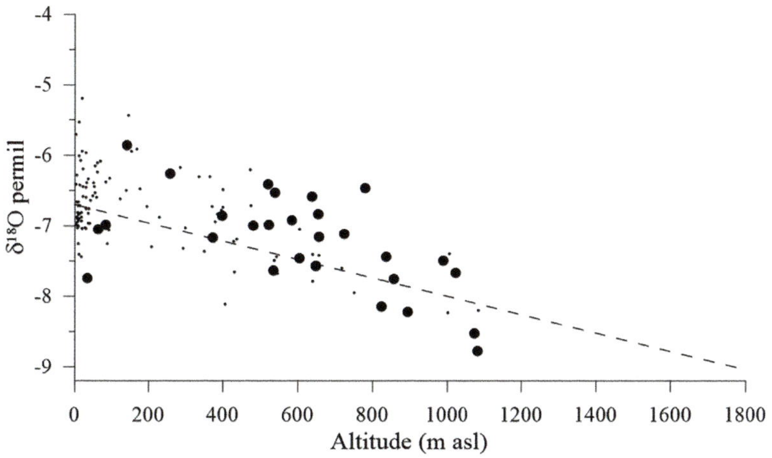

Fig. 7.7 Relationship between $\delta^{18}O$ and altitude

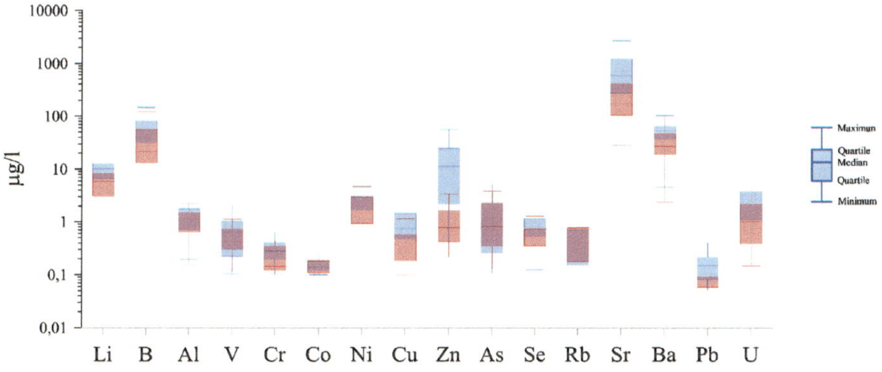

Fig. 7.8 Box-whisker plot of trace elements at Peloritani Mountains basin. Red box spring data (this work). Light blu box data from INGV (2005)

Bibliography and Recommended Further Reading

Amantia A, Barbagallo M, Bivona C, Cubito A, Ferrara V, Pappalardo S, Pennini A (1993) Summary of research and illustrative mapping of aquifer vulnerability to pollution in the eastern sector of the Peloritani Mountains (Northeastern Sicily). In: Proceedings 2nd International Conference Of Geohydrology, Firenze 29 novembre-3 dicembre 1993, in Environmental Protection Techniques Notebooks, 49,pp 239–246, Pitagora Editrice

Amodio Morelli L, Bonardi G, Colonna V, Dietrich D, Giunta G, Ippolito F, Liguori V, Lorenzoni F, Paglionico A, Perrone V, Picarretta G, Russo M, Scandone P, Zanettin-Lorenzoni E, Zuppetta A (1976) The calabrian-peloritanian arc in the apennine-maghrebide orogen. Mem Soc Geol It 17:1–60

Arnone G, De Rosa P, Mascari A (1979) Geological observations in the area of Longi (Western Peloritani Mountains). Boll Soc Geol It 98:217–226

Baldacci L (1886) Geological description of the island of Sicily. Mem Descr della Carta Geol d'It 1:408

Bonardi G, Giunta G, Liguori V, Perrone V, Russo M, Zuppetta A (1976) Geological scheme of the Peloritani Mountains. Boll Soc Geol It 95:49–74

Bonardi G, Giunta G, Perrone V, Russo M, Zuppetta A, Ciampo G (1980) Observations on the evolution of the Calabrian-Peloritanian arc in the Lower Miocene: the Stilo-Capo d'Orlando Formation. Boll Soc Geol It 99:365–393

Caire A, Duée G, Truillet R (1965) The Limestone Range of the Peloritan Mountains (Sicily). Bull Soc Géol France 7(7):881–888

Cangemi M, Censi V, Madonia P, Favara R (2021) Application of geostatistical tools to the geochemical characterization of the peloritani mts (Sicily, Italy) Aquifers. Water 13:3269. https://doi.org/10.3390/w13223269

Catalano R, D'Argenio B (1982) Geological scheme of Sicily. In: Catalano R, D'Argenio B (eds) Guide to the Geology of Western Sicily. Geological Guides

Catalano R, D'Argenio B (1978) An essay of palinspastic restoration across the Western Sicily. Geol Rom 17:145–159

Civita M (1972) Hydrogeological schematization of normal springs and related catchment works. Mem e Note Ist Geol Appl 12

Dongarrà G, Manno E, Sabatino G, Varrica D (2009) Geochemical characteristics of waters in mineralised areas of Peloritani Mountains (Sicily, Italy). Appl Geochem 24:900–914

Ferla P, Meli C (2007) Petrogenesis of tourmaline rocks associated with Fe–carbonate–graphite metapelite, metabasite and strata-bound polymetallic sulphide mineralization, Peloritani Mountains, Sicily, Southern Italy. Lithos 99:266–288

Ferrara V (1999) Vulnerability to pollution of the Peloritani area aquifers. Northeastern Sicily. Notebooks of Environmental Protection Techniques., 66. Pitagora Editrice, Bologna, pp 119

Gargano C (1994) Geologic map of Messina and the northeastern sector of the Peloritani Mountains (NE Sicily). Scale1:25.000. S.El.Ca., Firenze

Giunta G, Bellomo D, Carnemolla S, Runfola P (1992) Geological map of the western sector of the Longi-Taormina line, between S. Fratello, Longi, Monte Sollazzo (northeastern Sicily). Scale 1:50.000

Giunta G, Messina A, Bonardi G, Nigro F, Somma R, Cutrupia D, Giorgianni A, Sparacino V (1998) Geology of the Peloritani Mountains (NE Sicily). Hiking guide, 77° Summer meeting, Palermo, Settembre 1998

Istituto Nazionale di Geofisica e Vulcanologia (INGV), Sezione di Palermo (2005) First characterization and monitoring of Groundwater aimed at the drafting of the "Water Protection Plan of the Region of Sicily". On behalf of the Sicilian Region – Delegate Commissioner for the Waste Emergency and Water Protection

Lentini F, Catalano S, Carbone S (2000) Geological map of the province of Messina. Regional Province of Messina, Tip. Selca, Firenze

Liotta M, Grassa F, D'Alessandro W, Favara R, Gagliano Candela E, Pisciotta A, Scaletta C (2013) Isotopic composition of precipitation and groundwater in Sicily, Italy. Appl Geoch 34:199–206

Madonia P, Cangemi M, Favara R (2021) Modeling rain isotopic composition under orographic control: a landscape approach for hydrogeological applications. Hydrology 8:22. https://doi.org/10.3390/hydrology8010022

Ogniben L (1960) Explanatory note on the geological scheme of northeastern Sicily. Riv Min Sic II:183–212

Schmidt di Friedberg P (1963–64) Petroleum lithostratigraphy of Sicily. Riv Min Sic 15, 88–90:198–217

Wezel FC (1970) Geology of the numidian flysch of northeastern sicily. Mem Soc Geol It 9:225–280

Piazza Armerina Hydrogeological Basin

8

Leonardo La Pica, Giovannella Pecoraino, Esterina Gagliano Candela, and Sabina Morici

8.1 Geological Framework

The Piazza Armerina basin is located in central-eastern Sicily, including part of the territory of the provinces of Enna and Caltanissetta, bordered to the south by the Mediterranean Sea, to the southeast by the Hyblean plateau, to the east by the Catania Plain and the volcanic edifice of Etna, to the north by the Nebrodi Mts and to the west by the Caltanissetta Trench (Figs. 8.1, 8.2).

In this sector of Sicily, the Erei Mountains form a set of moderate reliefs and plateaus that run from the northern Nebrodi-Madonie chain towards S-SE to the Hyblean Plateau. These reliefs form a line of displuvium that crosses central Sicily, separating the catchment areas of the Ionian Sea from those of the Mediterranean.

The highest altitudes and the greatest differences in height of the Mts Erei are reached in its northern portion, from Gangi to Calascibetta at Mount Zìmmara (1333 m) and Mount Altesina (1.112 m). These greater morphological roughnesses are closely related to the lithological characteristics of the outcrops. In the area, in fact, large plates formed by powerful quartzarenitic bed alternating with reduced clayey intervals outcrop to form the Numidian Flysch unit (Ogniben 1960).

L. La Pica · G. Pecoraino (✉) · E. Gagliano Candela (Deceased) · S. Morici
Istituto Nazionale di Geofisica e Vulcanologia, Sezione di Palermo, Via Ugo La Malfa 153, 90146 Palermo, Italy
e-mail: giovannella.pecoraino@ingv.it

L. La Pica
e-mail: leonardo.lapica@ingv.it

S. Morici
e-mail: sabina.morici@ingv.it

E. Gagliano Candela et al. (eds.), *Drinking Water in Sicily*,
https://doi.org/10.1007/978-3-031-60473-7_8

Fig. 8.1 The Piazza Armerina hydrological basin. Imagine from Google Earth Pro v 7.3.6.9345 (December 29, 2022). Sicily Italy

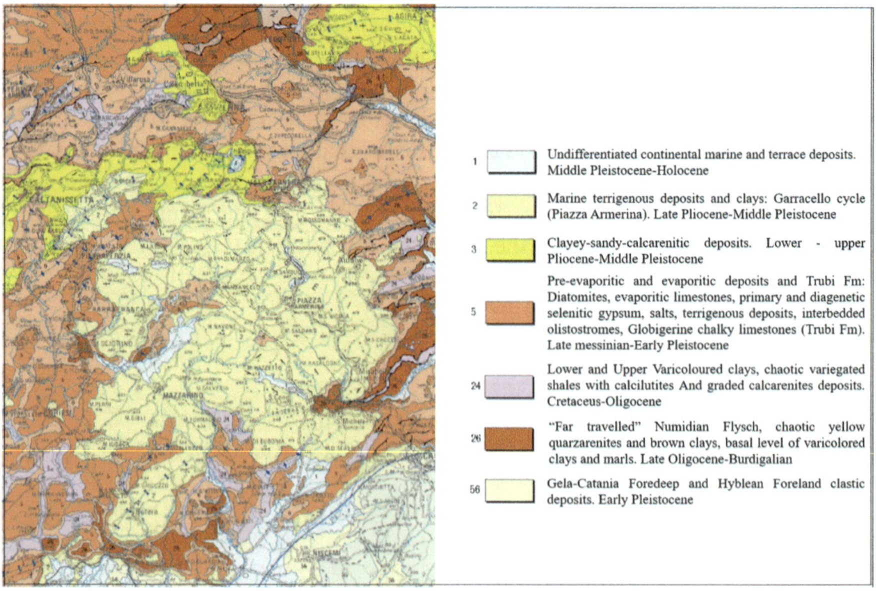

Fig. 8.2 Geological scheme of Piazza Armerina area (modified from Lentini and Carbone 1995)

More or less synclinal plates of marly and sandy arenitic soils of the Plio-Pleistocene age instead determine the modest reliefs of the southern part of the Mts Erei.

The area is part of the foreland domain known as the Caltanissetta Basin (Catalano and D'Argenio 1982). Actively subsiding during the Neogene and Quaternary periods, this basin, based on allochthonous units of the Sicilid Complex (Ogniben 1960), is filled by Mio-Pliocene and Pleistocene post-orogenic soils (Roda 1971). In reality, it represents a system of contiguous, syntectonic sedimentary basins that migrated in concomitance with the translation and shortening events that affected the Apennine-Maghrebid chain (Lentini et al. 1991).

The following lithological units can be recognised from bottom to top:

- Flaky Clays (Cretaceous-Eocene)
- Cozzo Terravecchia Unit (Upper Miocene—Lower Tortonian)
- Brecciated clays (A.B.II)
- Alternation of diatomite and snow-white marls ("Tripoli") (Upper Tortonian)
- Chalky-sulphur series (Upper Tortonian—Messinian)
- Globigerinid marly limestones ("Trubi"), with intercalations of A.B. IV (Lower Pliocene)
- Marly sandstone unit (Upper Pliocene—Lower Pleistocene)
- Blue clays and clayey marls with intercalations of brecciated clays (A.B.V)
- Clayey sands and sandy marly clays
- Quartz sands and calcarenites (Upper Pliocene)
- Ancient and recent alluvial deposits, sometimes terraced (Pleistocene)
- Lacustrine deposits (Pleistocene-Actual)
- Landslides, debris and landfills (Current)

Flaky Clays

The flaky clays are the oldest soils outcropping in the basin and lithologically consist of a rather tectonised clay mixture, including blocks of various lithologies (limestone with rudists, basaltic blocks, limestone with foraminifera, etc.) up to hundreds of m^3 in size, referred to in geological literature as 'exotics'. Sediments belonging to the Sicilidi units and soils of the Numidian Flysch contribute to the unit of the flaky clays.

Cozzo Terravecchia Unit

This unit is composed of fluvial-deltaceous deposits followed towards the south by pelitic and then pelitic-biogenic deposits. The typical locality of the unit is the northern sector of Cozzo Terravecchia, north of S. Caterina di Villaermosa (Schmidt di Friedberg 1962). The high Miocene soils of Baldacci (1886), the Tortonian-Helvetian clayey marls of the Licata Unit (Ogniben 1954), the Tortonian olistostromes and the high Tortonian detrital organic limestones can be traced back to this unit.

Blasted Clays (A.B.)

These clays are found interspersed with normal Tortonian sedimentation. They are composed of clayey and marly fragments, up to a few cm in size, characterised by a microfauna that can be traced back to the Tortonian period, immersed in a clayey matrix with Oligocene microfauna. The maximum thicknesses are found in the syncline cores, and the minimums in the anticline cores. Five levels of A.B. have been distinguished:

- A.B. I: intercalated with the Tortonian sediments
- A.B. II: immediately underlying the gypsum-sulphur series
- A.B. III: intercalated with the chalky-sulphur series
- A.B. IV: intercalated with the Trubi
- A.B. V: intercalated with the overlying Pliocene soils

In the study area, there are small levels attributable to A.B. II and A.B. V.

"Tripoli" (Upper Tortonian)

The "Tripoli" lithological unit is composed of an alternation of diatomite and snow-white marls. The depth of sedimentation is only a few tens of metres (max 60 m); the environment is euxinic (Ogniben 1957). It emerges in rather limited thicknesses and is often covered by agricultural soil or groundwater debris.

Chalk-Sulphur Series (Upper Tortonian—Messinian)

Chalks of the Second Cycle (Pasquasia Chalks)

The Pasquasia Chalks are an alternation of Balatino and selenitic gypsum with clayey marls. In the northern areas of Sicily there is an increase in the terrigenous content, which further north becomes prevalent on the chalks.

Arenazzolo

This lithological unit consists of a clastic horizon with an arcosic composition, paleontologically almost sterile, and limited thickness.

Chalks of the First Cycle (Cattolica Eraclea Chalks)

These chalks insist of banks up to 20 m thick, often dismembered due to the intramessinian tectonic phase. They constitute a heteropic facies of the basic limestone.

Basal Limestone

This rock series is formed by alternating 1–2 m thick banks of predominantly evaporitic limestone, and subordinately by marls and breccias. There is no clear transition with the Tripoli, but there are alternations of the two terms. The unit is paleontologically sterile.

Globigerinid Marly Limestones (Trubi, Lower Pliocene)
They consist of marls and calcareous marls with a $CaCO_3$ content ranging from 40 to 85%. Outcrops are found as alternating marls and clayey marls in layers of 20–80 cm. The layer stratification is transgressive and generally discordant on the upper Miocene soils. The average thickness of the unit is 100 m.

Marly-Arenaceous Unit (Upper Pliocene - Lower Pleistocene)

Grey-Blue Clayey Marls, Clays and Marly Clays with Intercalations of Brecciated Clays (A.B.V)
These sedimentary deposits are generally found discordant on the Trubi, with a clear transition between the two terms. The clay content generally increases towards the top where marly clays are present. The presence of Globorotalia bononiensis, Globorotalia hirsuta and Globorotalia crassaformis makes it possible to date the marly portion to the Lower and Middle Pliocene and the strictly clayey portion to the Upper Pliocene. They vary in thickness between 100 and 200 m. At the top, the lithology changes very gradually with an increase in grain size, until the clays pass into the more recent sands.

Clayey Sands and Sandy Marly Clays
This series represent the continuation of the regressive facies initiated in the underlying soils. They consist of alternating banks of fine sand with banks of silty sand and siltstones, and also have interspersed arenitic lenses. They have a maximum thickness of about 250 m.

Quartz Sands and Calcarenites
This is the most extensive unit in the study area, consisting of a thick series of yellowish quartz sands interbedded with numerous cemented levels with a highly variable lateral extension, ranging from about ten metres to as much as 4–5 km. Its transgressive layering is clearly evident in some places and is manifested by marked angular discordances. The micro- and macro-fauna are indicative of a littoral environment with a predominance of coastal or shallow forms.

Ancient and Recent Alluvial Deposits (Sometimes Terraced)
The alluvial deposits all have a limited extension; they are small terraces located up to 30 m above the current level of the watercourses. They consist of poorly graded banks of quartzarenitic pebbles, generally rather coiled and with indications of flattening. Thickness of the deposits is around a couple of metres.

Recent gravelly-sandy alluvium can be found with a thickness of several metres, occasionally less, along the watercourses.

Two types of structures can be distinguished in the area, differing essentially in the intensity of the folding. The first type involves pre-evaporitic and evaporitic soils that have been strongly folded by the same act of diastrophism, while the second type can be observed in marly and sandy soils. They differ because the second type gives rise to milder dislocations.

Table 8.1 Scheme of water body in the Piazza Armerina hydrological basin

Water body	Hydrogeological complex	Mouton hydrogeological complex	Description	PTA hydrogeological complex	Type of aquifer
Piazza Armerina	Piazza Armerina	DET plio-quaternary debris formation	Sandy complex	Yellowish, well stratified limestones in levels varying in thickness from a few centimetres to a few mwtres, with frequent fossiliferous horizons containing oligotupic ostreid and pectinid faunas and clayey sands and silts in level of various thicknesses	Multilayer aquifer

The two structural types are the consequence of an energetic tectonic phase that affects pre-evaporitic and evaporitic soils and separates them from two successive sedimentary cycles between the Upper Pliocene and Sicilian periods. It is preceded by the important Tortonian orogenic phase of regional character.

The thick sandy accumulations, which distinguish the sandy-quartzarenitic complex found on the outcrop, are interpreted as the frontal beds of a large delta advancing towards the southern coast, fed by material from the dismantling of the Nebrodi-Madonie area to the north.

From a structural point of view, the soils of the two sedimentary cycles are considered post-orogenous with respect to the folded stratum acme of the Upper Pliocene that affected the pre-evaporitic and evaporitic soils; they are in fact characterised by large syncline structures, slightly dislocated, free from disjunctive tectonic events.

In the Piazza Armerina hydrological basin only one water body has been distinguished (Table 8.1).

8.2 Hydrogeology

Outcropping terrains in the area were divided into different permeability classes:

1. Very low permeability or impermeable;
2. Localised medium permeability;
3. high to medium diffuse permeability.

The first class includes predominantly clayey terrains in which there is no significant water circulation. The second class includes marly terms and alternating clays and quartzarenites in which there is limited water circulation in the more carbonate levels and in the more fissured arenaceous levels.

The terrains belonging to the third class make up the aquifers, which take on different significance in relation to the quantity and quality of the resources they contain.

The terms that have the greatest significance from a hydrogeological point of view are the sandy and calcarenitic ones. As a whole, they tend to be of higher permeability, which in the sands is medium to high due to porosity, although with marked variability, both horizontally and vertically, due to the presence of clayey-sandy-silt intercalations. In the cemented levels, permeability is generally high for cracking, and to a lesser extent for porosity. These conditions, which imply frequent permeability limits, often indefinite, within the complex, favour the existence, in addition to suspended and sometimes overlapping water tables, of a large number of springs of modest productive significance (Coltro et al. 1984).

In alluvial deposits, both current and recent, the permeability varies considerably in relation to the percentage of fine fraction present; it remains generally low, but may be locally higher due to the increase in the sandy and/or pebbly component.

In the evaporitic terrains, which are permeable due to cracking, but also due to porosity in the case of brecciated and vacuolar limestone, water circulation is discontinuous due to the presence of marly and clayey levels between the limestone and gypsum banks.

The following hydrogeological complexes have been distinguished (Fig. 8.3):

Sandy-Calcarenitic Complex

This complex is made up of a group of plates of variable extension, resting on marly and clayey terms of various ages. They present a high permeability but the water circulation is complex due to the frequent variations in permeability, both vertical and horizontal, caused by the presence of clayey-sandy levels of variable size and thickness.

In this complex there are superficial, intermediate and deep aquifers. The former consist essentially of very discontinuous aquifer levels, located at a modest depth from ground level and conditioned by meteoric inflows. They are partly exploited by means of large-diameter wells and give rise to ephemeral springs.

The intermediate aquifers affect the middle-upper portion of the complex, being quite persistent in time, but discontinuous in space. This is due to the presence of poorly permeable levels that fragment the water circulation, also being responsible for local phenomena of semi- or total confinement. The productivity of these aquifers can be interesting, especially in cases where they are under pressure. The deep aquifer is the main source of seepage water and is characterised by considerable productivity. It rests on an impermeable substratum represented by marly

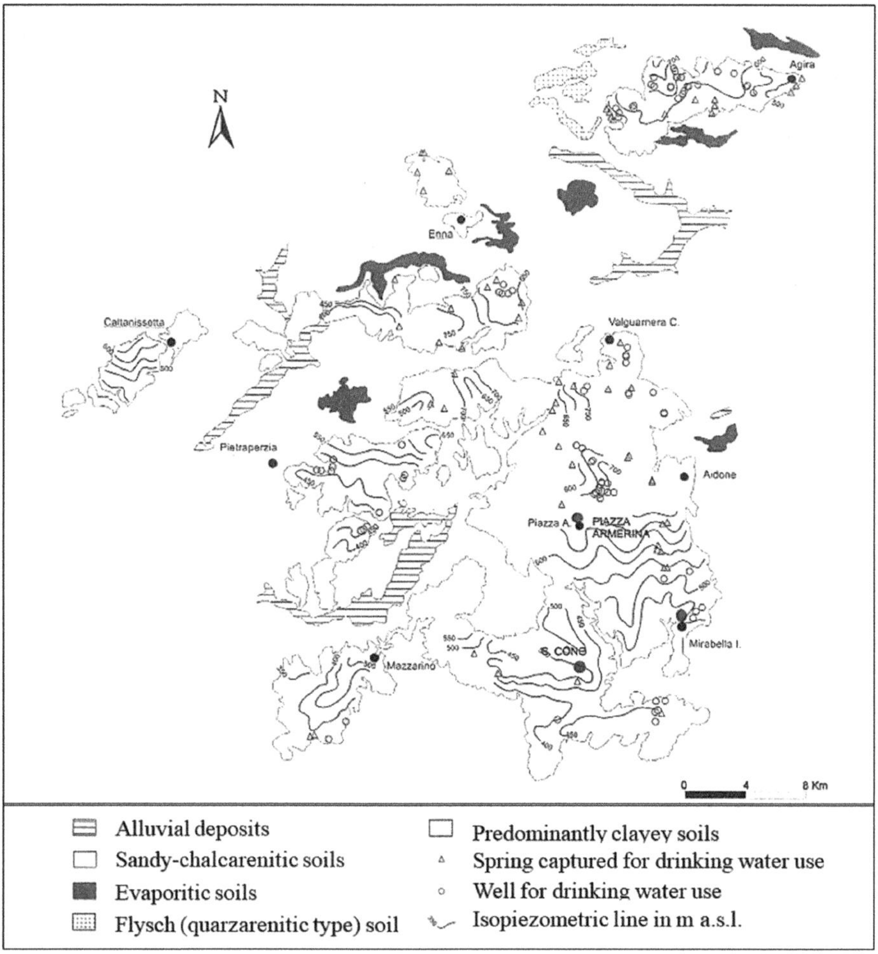

Fig. 8.3 Schematic hydrogeological map of the study area (modified from Ferrara et al. 1999)

and clayey soils of various ages, whose structure conditions the direction of underground flow, which occurs at the contact between the aquifer and the substratum at the lowest points.

Evaporitic Complex

The evaporitic complex is discontinuous and of modest lateral extent and contains water of poor quality due to excess sulphates. It is not significant for hydrogeological purposes.

Alluvial Complex

Its thickness is generally limited and its permeability tends to be low and varies from point to point; the water circulation is fragmented, giving rise to water tables with modest productivity and a prevalently seasonal character.

The complex of calcarenites, sands and siltstones, dating from the Middle Pliocene and the Lower Pleistocene, represents the unit of greatest hydrogeological interest in the entire examined area. It is made up of yellowish calcarenites, well stratified, in levels varying in thickness from a few centimetres to a few metres, with frequent fossiliferous horizons, containing oligotypic ostreid and pectinid faunas, and of clayey sands and silts, in levels of varying thickness.

Water resources in the area are mainly exploited by means of drilled wells with a depth of between 50 and 200 m and a productivity that varies from fractions of l/s to a maximum of 20 l/s. The wells are generally not equipped with effective protection areas and often present problems of functionality due to clogging and accelerated wear of the lifting systems due to constructional deficiencies. There is a greater concentration of wells, most of which have high productivity, to the north of the built-up area of Piazza Armerina; these wells are used to supply drinking water to the built-up area and other urban centres.

The springs are tapped by means of intakes or tunnels which are generally of modest length; average productivity is never significant and reaches a few l/s at most.

The calcarenitic sandy complex has been attributed a high to medium degree of vulnerability, being characterised by:

- High vertical and horizontal permeability at the arenaceous levels where infiltration and propagation of pollutants is easier;
- High persistence times within the sandy component, which leads to unfavourable conditions in the case of non-degradable pollutants;
- Absence of sufficient cover with reduced permeability;
- Easy interchange between surface water and groundwater along valley incisions;
- Often low water table submergence.

The centres of danger are essentially represented by built-up areas and agricultural and livestock farming activities, while industrial activities, which are scarcely present in the area, definitely play a marginal role.

Almost all inhabited areas do not have efficient sewage systems or waste water treatment plants, and therefore waste water easily infiltrates the permeable sandy-calcite soil and is a potential source of pollution. In addition, there are uncontrolled MSW dumps and abandoned quarries where various types of waste are dumped.

Exposure to pollution risk is limited according to the type and distribution of potential pollutant producers, given the type of activities in the area and the prevailing land use. The risk is higher in the areas surrounding population centres due to the lack of waste disposal infrastructures.

8.3 Hydrogeochemical Characterisation

Due to the geological characteristics of the hydrogeological basin under study, there is only one significant water body in which all springs and wells for drinking water use are located. Three springs were sampled in this study (Fig. 8.4).

Again, literature data were considered for completeness. Considering all available samples, in the Langelier–Ludwig classification diagram (Fig. 8.5), the waters

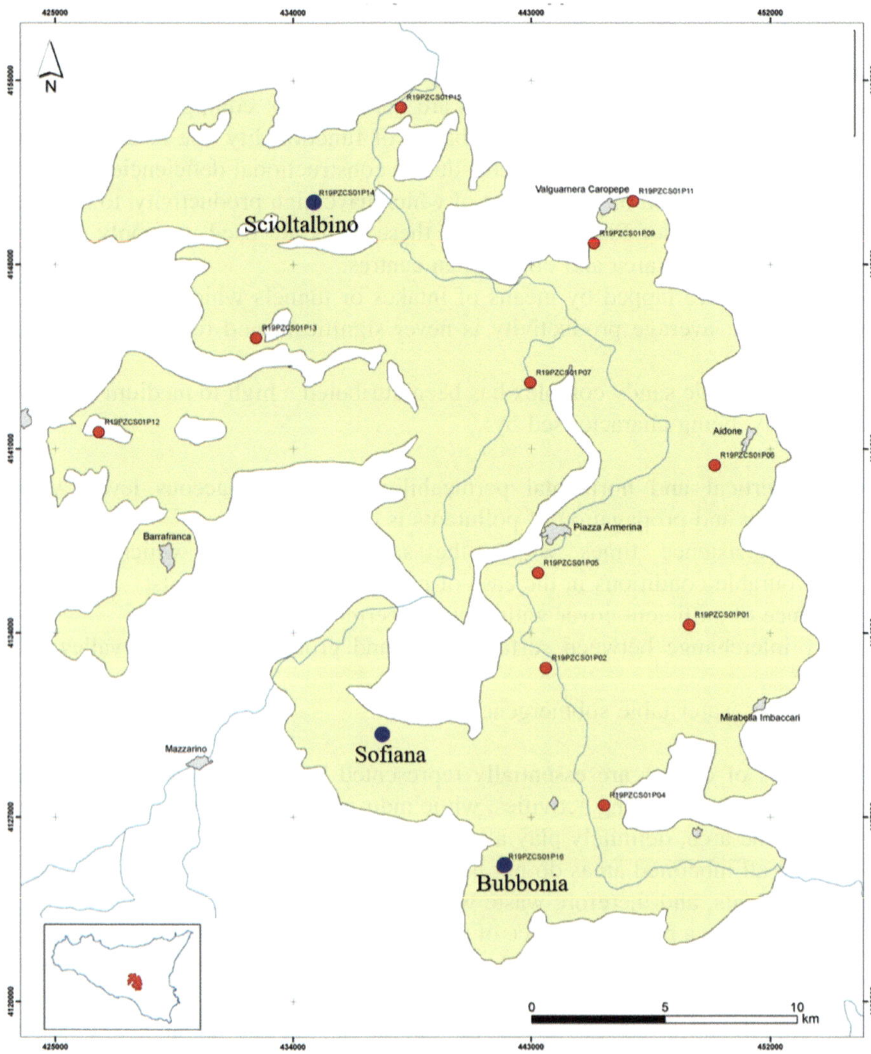

Fig. 8.4 Springs for drinking water use in the Piazza Armerina. Blue dots this study, red dots from INGV (2005)

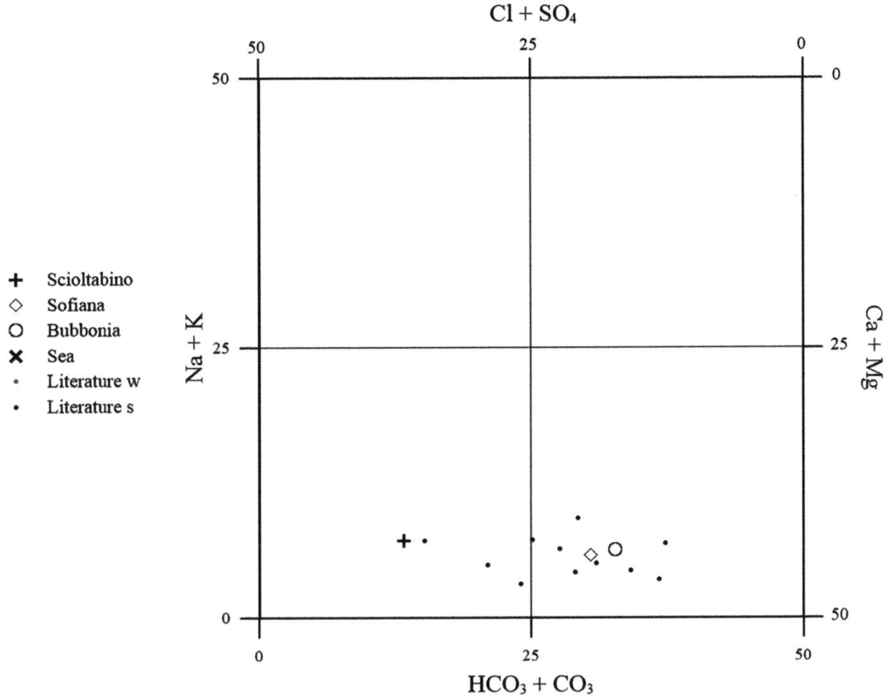

Fig. 8.5 Langelier–Ludwig classification diagram. Springs fall in the bicarbonate-alkaline-earthy, chlorine–sulphate–alkaline-earthy quadrants. Literature data (La Pica et al. 2020) are shown for comparison; S: spring, W: well

show a trend from bicarbonate-alkaline-earthy to chloride–sulphate–alkaline-earthy.

This trend is also recognisable in Fig. 8.6 in which variations towards the vertices of SO_4 and, to a lesser extent Ca, are evident.

The increase in SO_4 may be a consequence of the interaction of groundwater with the rocks of the "Gessoso Solfifera serie" underlying the calcarenites.

These considerations suggest that they are in the presence of a non-homogeneous body of water and thus represented by geochemically different waters.

8.4 Isotopic Characteristics of the Catchment Area

The Piazza Armerina basin is located in central Sicily, the most arid area of the region. The isotopic composition of groundwater show a very tight range: from −7 to −6‰ for $\delta^{18}O$ and from −37 to −31‰ for δ^2H (Table 8.2). Sampled spring waters align on the LMWL for the central Sicily (Liotta et al. 2013) as sector 2 (Table 1.8). The LMWL is $\delta^2H = 7.67\ \delta^{18}O + 13.9$.

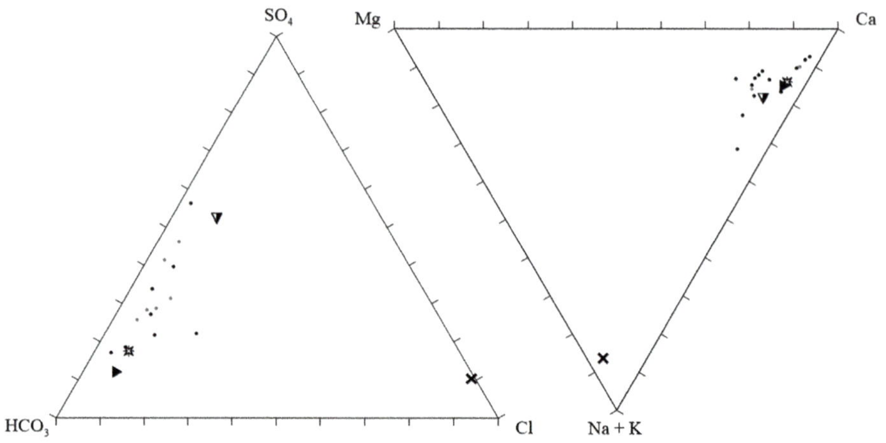

Fig. 8.6 Anions (left) and cations (right) ternary diagrams for Piazza Armerina catchment area. Literature data are shown (La Pica et al. 2020). S: spring, W: well

Table 8.2 Chemical and isotopic composition of the sampled waters

	T	EC	pH	Eh	Na	K	Ca	Mg	Cl
Scioltabino	16.8	761	7.75	196	1.08	0.209	6.67	1.13	1.17
Bubbonia	16.2	573	7.21	154	0.84	0.032	5.6	0.51	0.79
Sofiana	17.2	173	7.16	129	0.76	0.054	5.74	0.51	0.77

	NO_3	SO_4	HCO_3	δ^2H	$\delta^{18}O$	WB
Scioltabino	1.14	4.3	2.41	− 35	− 6.28	PA
Bubbonia	0.77	0.89	4.65	− 33	− 6.01	PA
Sofiana	0.7	1.28	4.28	− 32	− 6.11	PA

Analytical data are in meq/L, EC in μS/cm and δ^2H and $\delta^{18}O$ in ‰ versus SMOW. WB is for water body. PA is for Piazza Armerina water body

Collected and literature data plot between the GMWL and MMWL lines, at the local meteoric water line, indicating a direct meteoric water supply and, a rather shallow and short circulation (Fig. 8.7).

Applying the isotopic gradient relative to sector 2 of 0.16 δ‰/100 m (Table 1.7) from Liotta et al. (2013), all sampled waters stay above the line representing the isotopic gradient suggesting the evaporation of rainwater drops just in the soil before reaching the aquifer (Fig. 8.8). Climatological studies highlight a strong soil potential evapotranspiration such that water deficit conditions (rainwater < evaporated water) occur on average 6–7 months per year.

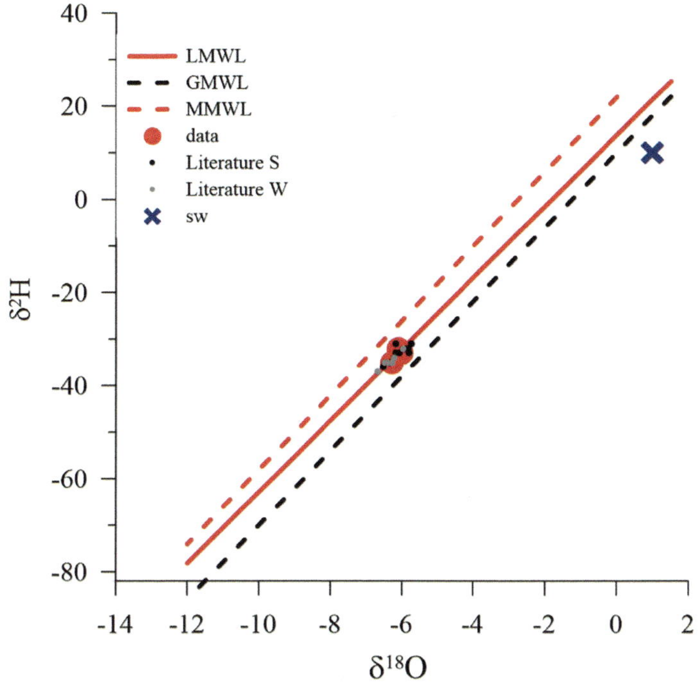

Fig. 8.7 δ^2H versus $\delta^{18}O$ (in ‰ versus SMOW) diagram of spring waters. The LMWL is from Liotta et al. (2013). The GMWL and MMWL are shown for comparison. Literature data from La Pica et al. (2020). S: spring, W: well

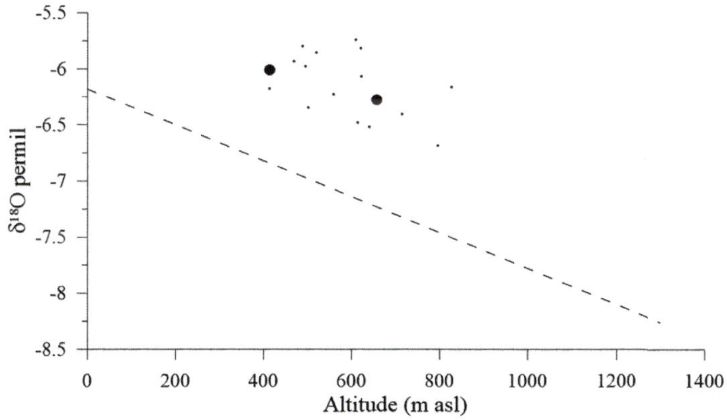

Fig. 8.8 Relationship between $\delta^{18}O$ and altitude

Bibliography and Recommended Further Reading

Baldacci G (1886) Geological description of the island of Sicily. Description of the Geological Map of Italy, Roma

Bello M, Franchino A, Merlini S (2000) Structural model of Eastern Sicily. Mem Soc Geol It 55:61–70

Catalano R, D'Argenio B (1982) Geological outline of Sicily. Guide to the geology of western Sicily., pp 9–36

Coltro R, Ferrara V, Risita A (1984) Contribution to the knowledge of the groundwater resources of Ennese-Nisseno (Central Sicily). Consortium of drainage of 2nd degree of the province of Enna. Notebook n. 1.

Craig H (1961) Isotopic vaiations in meteoric waters. Science 133:1702–1703

Ferrara V, Maugeri S, Pappalardo G (1999) The aquifers of the central-eastern area of Sicily: water resources, water quality and vulnerability to pollution. Proceedings of the 3rd National Conference on Underground Water Protection and Management for the Third Millennium. Parma, 13–14–15 october 1999

Gat JR, Carmi I (1970) Evolution of isotopic of atmospheric waters in the Mediterranean sea area. J Geophis Res 75:1437–1440

Istituto Nazionale di Geofisica e Vulcanologia (INGV), Sezione di Palermo (2005) First characterization and monitoring of groundwater aimed at the drafting of the "Water Protection Plan of the Sicily Region". On behalf of the Sicilian Region – Deputy Commissioner for the waste emergency and water protection

La Pica L, Morici S, Favara R, Gagliano Candela E, Pecoraino G, Scaletta C (2020) Chemical-physical parameters, major, minor and trace elements composition of groundwater belonging to the hydrogeological basin of the Piazza Armerina in Sicily (Italy). Interdiscip Earth Data Alliance (IEDA). https://doi.org/10.26022/IEDA/111746

Lentini F, Carbone S, Catalano S, Grasso M, Monaco C (1991) Presentation of the geological map of east-central Sicily. Mem Soc Geol It 47:145–156

Lentini F, Carbone S (1995) Geological map of Sicily. Mem Soc Geol It

Liotta M, Grassa F, D'Alessandro W, Favara R, Gagliano Candela E, Pisciotta A, Scaletta C (2013) Isotopic composition of precipitation and groundwater in Sicily, Italy. Appl Geoch 34:199–206

Ogniben L (1953) "Argille Scagliose" and "Argille Brecciate" in sicily. Boll Soc Geol It 75:281–289

Ogniben L (1957) Petrography of the chalky sulfiferous series and related geological considerations. Mem Descr Carta Geol It 33:251–261

Ogniben L (1960) An illustrative note of the geological scheme of northeastern Sicily. Riv Min Sic 64–65:183–212

Ogniben L (1954) Sicilian Brecciated Clays with detailed reliefs of Grottacalda-Zubbi-Passarello. Mem Ist Geol Min Univ Padova, vol 18

Roda C (1968) Geology of the Pietraperzia tablet (Caltanissetta and Enna Provinces, sheet 268, III NE). Atti Acc Gioenia Sc Nat 19:145–254

Roda C (1971) Illustrative note to the geological map of the M. Gibiliscemi tablet (Province of Caltanissetta, Sheet. 272, I SW). Boll Acc Gioenia Sc Nat 4(10):571–632

Schimdt di Friedberg P (1964) Petroleum lithostratigraphy of Sicily. Riv Min Sic 15:198–217

Wezel FC (1966) News about the highly detailed geological survey of the Mirabella Imbaccari tablet (Sheet 272, I NE). Boll Acc Gioenia Sc Nat 8:652–657

Wezel FC (1965) Geology of the Mirabella Imbaccari tablet (Prov. Catania, Caltanissetta and Enna, Sheet 272, I NE). Boll Soc Geol It 84:3–136

Rocca Busambra Hydrogeological Basin

9

Sabina Morici, Giovannella Pecoraino, Esterina Gagliano Candela, and Leonardo La Pica

9.1 Geological Framework

Located in the central-western part of Sicily, the Rocca Busambra hydrogeological basin covers an area of approximately 140 km^2 in the province of Palermo, in the municipalities of Corleone, Godrano, Cefalà Diana, Villafrati, Monreale, Campofelice di Fitalia and Mezzojuso (Fig. 9.1).

The geomorphological configuration of the area, reflecting the lithologies of the outcropping terrains, appears very varied and complex. We pass, in fact, from rather soft morphologies in the slightly sloping slopes, set on plastic soils, such as clays and sandy clays, to steep slopes with sharp gradient breaks in the carbonate reliefs. The carbonate complex of the Rocca Busambra ridge towers over the entire territory. The ridge, made up of several hills (Pizzo Nicolosi, Rocca Argenteria, Rocca Ramusa and Pizzo Busambra), reaches an altitude of 1.613 m a.s.l. Even in the quartzarenitic soils of the Numidian Flysch, outcropping to the north of Rocca Busambra, there is a very rugged landscape, characterised by steep slopes, juxtaposed with significant incisions (Contrada Cerasa) and gullies along the southern side (Contrada Lavanche). Karst dissolution has acted on part of the carbonate rocks, forming, in the western part of the area, along the Frattina river, the 'Gole del Drago' (Fig. 9.2).

S. Morici · G. Pecoraino (✉) · E. Gagliano Candela (Deceased) · L. La Pica
Istituto Nazionale di Geofisica e Vulcanologia, Sezione Di Palermo, Via Ugo La Malfa 153, 90146 Palermo, Italy
e-mail: giovannella.pecoraino@ingv.it

S. Morici
e-mail: sabina.morici@ingv.it

L. La Pica
e-mail: leonardo.lapica@ingv.it

E. Gagliano Candela et al. (eds.), *Drinking Water in Sicily*,
https://doi.org/10.1007/978-3-031-60473-7_9

Fig. 9.1 The Rocca Busambra hydrological basin. Image from Google Earth Pro v 7.3.6.9345 (December 29, 2022). Sicily, Italy

Rocca Busambra constitutes the natural watershed of three of the most significant hydrographic basins in western Sicily: the Belice Sinistro, the San Leonardo and the Eleuterio.

The study area has dendritic drainage in correspondence with the plastic soils outcropping at the base of the Rocca Busambra ridge.

Numerous torrential channels originate at the foot of the great massif and merge on the clayey-sandstone slopes to form the "Torrente Scanzano" on the northern side, a tributary of the Eleuterio. The river, dammed in 1962, feeds the Scanzano artificial reservoir.

From the slopes of Pizzo Nicolosi and Rocca Ramusa branch off the waters of the Frattina torrent which, joining with the Catagnano torrent coming from Mt Leardo to north, give rise to the Belice Sinistro. This one, more properly called Fiume di Frattina, winds among the calcareous bastions of the Rocche di Rao, continuing southwards, and collecting the waters coming from the slope of Rocca Busambra. Two of San Leonardo's main tributaries also originate from its slopes: the Mendola River, which collects the waters of the Guddemi Valley and the Azzirolo Stream. The Guddemi valley originates on the south-eastern slope, where it carves out clay and marl units, drawing large gullies; the Azzirolo stream, on the other hand, furrows the wide valley dominated by Godrano and Mezzojuso.

Another main watershed is the Serra Capezzana which, located northeast of Busambra, marks the boundary between the San Leonardo basin and the Milicia

Fig. 9.2 Details of the Gole del Drago

river basin. The Buffa and Cefalà streams rise from this ridge; they furrow the base of Pizzo Chiarastella to join, in a northerly direction, the Mulinazzo stream.

The outcrops in the area can be grouped into different tectonic units, which derive from Miocene deformation of pre-existing palaeogeographic domains. Figure 9.3 shows the structural layout of the area.

Starting from the geometrically lower (and therefore less deformed) ones, the following units can be distinguished:

- Mufara Unit, which derives from the deformation of part of the Imerese–Sicana succession;
- Rocca Busambra Unit, which derives from the deformation of part of the Hyblean-Pelagian succession;
- Numidian Flysch Unit;
- Late-orogenic soils;
- Recent Cover materials.

Mufara Unit

Clay deposits (Carnic)

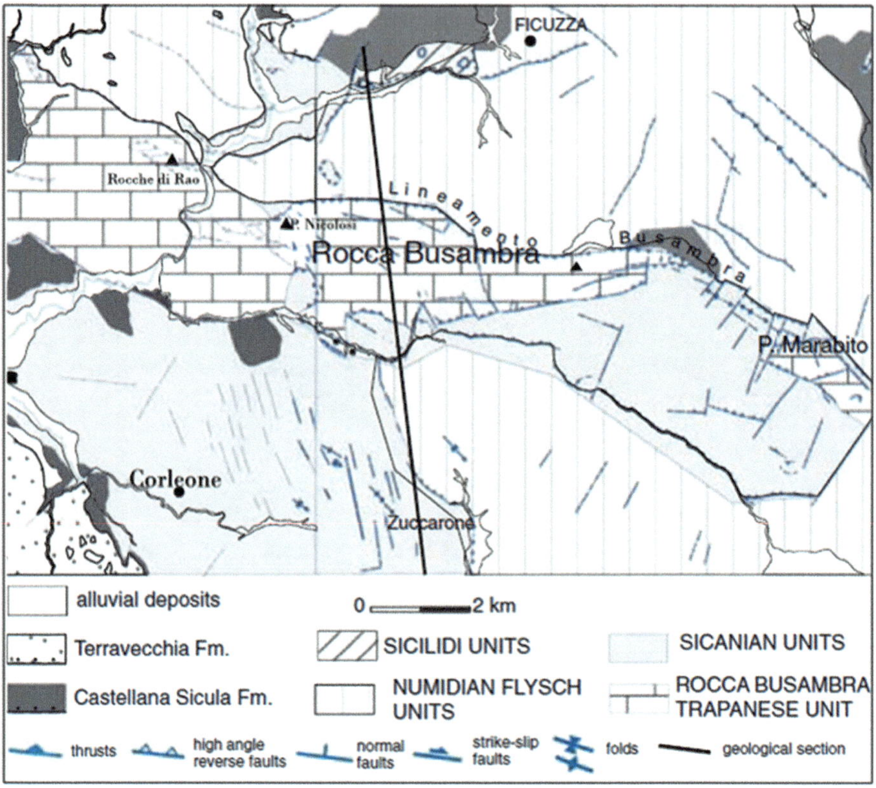

Fig. 9.3 Geological map of Rocca Busambra hydrogeological basin (modified from Basilone 2009)

Consisting of dark grey marly clays and calcilutites with radiolarians and lamellibranchs (Halobia and Daonella), these deposits represent the oldest lithological term present in the area. These soils, known in geological literature as the Mufara Unit, although sometimes improperly referred to as "Flysch Carnico", are dated to the Upper Trias. Within the area under consideration, these deposits outcrop in Contrada Pirrello (south of Cozzo Meriggio) and in Contrada Guddemi (south of Portella Guddemi), in an anomalous position; in fact, they are associated, through overthrusting, with the Numidian Flysch and the Marne di San Cipirrello, forming klippen.

Rocca Busambra Unit
This unit constitutes the entire Rocca Busambra dorsal. Some campaign data collected by Catalano et al. (1998), indicate that the ridge consists of at least two overlapping geological bodies with different stratigraphic characters, which can be considered as two distinct units:

In the typical succession, the following deposits can be recognized, from bottom to top:

Carbonate Deposits (Trias–Jura).

- *Platform limestones* (Upper Trias–Lower Lias), consisting of white oolitic and loferitic limestones, from massive to stratified in large banks, with typical bird's eyes structures and remains of algae (Cayeuxia, Dasycladacee and Paleodasycladus Mediterraneus), brachiopods, lamellibranchs and foraminifera (Valvulinides and Textulariides). They constitute the main body of the Rocca Busambra ridge and the surrounding reliefs: Rocca Argenteria, Rocca Drago, the Rocche di Rao, Monte Galiello (moved to the west) and Pizzo Marabito (moved to the east).
- *Condensed level* (Toarcian–Baiocian), consisting of beige-reddish limestone, impregnated with iron-magnesiferous crusts and with a rich fauna of ammonites, echinoderms, lamellibranchs and foraminifera.
- *Red ammonite and radiolarian limestones* (Upper Dogger–Lower Malm).
- *Reddish nodular limestones* (Lower Malm–Titonic), consisting of limestones with a rich fauna of Saccocoma and Globochaete Alpina, passing upwards to Calpionelle calcilutites.

Deep pelagic deposits (Cretaceous–Eocene)

- Crystalline limestone passing upwards to porous white marly limestone, with a conchoid fracture, known in geological literature as the Alcamo Unit or the Busambra Member, or by Sicilian quarrymen as Lattimusa (Upper Titonic–Neocomian), Alcamo Unit or Busambra Member, or by Sicilian quarrymen as *Lattimusa* (Upper Titonic–Neocomian).
- Marly limestones and calcareous marls, white to greenish-grey in colour, with concoid fractures, belonging to the Hybla Unit (Middle Cretaceous).
- Microcrystalline calcilutites, characterised by the presence of planktonic foraminifers immersed in a micritic matrix, belonging to the Amerillo Unit, also known as Scaglia Auct. (Upper Cretaceous–Eocene). These soils appear in a single large outcrop from Contrada Marosa to Piano Guddemi (although with different sedimentological characteristics, not recognizable in the field), and in small outcrops near Borgo Schirò, at Cozzo Marraccia, north of Piano Pilato, north of Masseria Pirrello and as a filling of the canyon of Pizzo Nicolosi, consisting of a small graben. Dividing the Cretaceous and Eocene Scaglia is the presence of carbonate Megabreccia (Maastrichtian).

In discordance, with a stratigraphic gap spanning the entire Oligocene, the next deposits follow:

- *Glauconitic calcarenites (*Aquitanian–Langhiano), consisting of strongly glauconitic calcarenites of a greenish-grey colour, rich in echinoderms, lamellibranchs and foraminifera fauna, known as *Corleone calcarenites* (Ruggieri 1966). They occur in small outcrops near contrada Casale and Cozzo Cavallo, with a tabular stratification, with levels of a few centimetres alternating with glauconitic sandy marls.
- Clayey marls (Middle–Upper Miocene), consisting of greenish-grey clayey marls with glauconite granules and arenaceous levels with planktonic foraminifera. Known as Marne di San Cipirrello, they occur with large outcrops in the central-western sector of the examined territory.

Numidian Flysch Unit
This unit consists of tobacco-coloured silty clays and argillites with intercalations of quartzarenitic and quartzoruditic lenses. The soils of this unit, which can be dated to the Oligocene- Lower Miocene, are difficult to interpret because of the different sedimentological characteristics with which it occurs in the various areas. The quartzarenitic intercalations occur in large banks with variable grain size, increasing from medium-fine in the lower portion to medium-coarse in the upper portion.

Late-Orogenic Units
These clays and marly clays, often with an abundant sandy fraction, ash-grey to bluish-grey in colour, are intercalated by large arenitic and conglomeratic beds. These rocks are rich in microfossils and foraminifera (Globigerina and Globorotalie), discordantly posed on the variegated clays, on the Numidian Flysch and on the grey marls of the Tortonian. They are part of the. Terravecchia Unit and are dated Upper Tortonian–Lower Messinian. The late-orogenic units occur in outcrops in the Tagliavia district, near the Rosario Sanctuary.

Recent Cover Materials

- Old and recent floods: these deposits, mostly clayey silt, form the bottom and current bed of the main watercourses.
- *Groundwater debris:* these are clastic deposits, consisting of loose or poorly cemented material, with carbonate fragments, lying along the main rocky slopes. They are the result of erosion and degradation of the rocks by exogenous agents. These detrital deposits are found all along the carbonate slope of the Busambra.

The *Rocca Busambra ridge*, in the model of Catalano et al. (1995), is located in the internal tectonic element of the Sicilian chain, precisely within the Internal Carbonate System.

In the area, Catalano et al. (1998) recognise:

- a Meso-Cenozoic carbonate body with the characteristics of a carbonate platform, structured in ramps with a general southern gradient and characterised by retrovergent shear planes;
- a lithological body with characteristics of bacinal carbonate facies, which appears internally deformed according to tectonic scales. In the southern sector, this seismic horizon seems to be correlated with the Cretaceous-Eocene bodies in flake facies outcropping at Portella del Vento. In the westernmost sector, where the Rocca Busambra unit depresses and disappears near Rocche di Rao, the carbonate seismic horizon of the basin appears more clearly continuous with the Sicanian successions that outcrop in the areas immediately to the south; this continuity would confirm the hypothesis of a possible belonging to the Sicanian domain of the deposits identified in the subsoil;
- a thick layer of clay-arenaceous and quartzarenitic deposits of the Numidian Flysch that overlaps the body of the basin deposits according to a flat surface and with internal imbrications of a lower order. The original overthrust contact between the Numidian Flysch Units and the underlying deposits is recognisable in the region.

The Rocca Busambra hydrological basin is subdivided into three water bodies:

Mezzojuso: constituited by the Numidian Flysh overlying the Mufara Un and Terravecchia Un. It is constituted by clays and silted clay with quarzoarenitic and quarzoruditic intercalations. The quartzarenitic interlayers occur in large banks with variable grain size, increasing from medium-fine in the lower portion to medium-coarse in the upper portion. The outcropping 'Numidian' rock successions consist of, from bottom to top:

- clays with frequent levels of fine sandstones with powerful turbiditic intercalations (Upper Oligocene - Lower Miocene);
- sandy clays and turbiditic sandstones (Lower Burdigean - Lower Langhien).

The permeability of the water body is caused by cracking within the fissured arenaceous beds. The water table is discontinuous, phreatic, but more often confined due to the rhythmic alternation of permeable quartzarenitic horizons and impermeable argillitic horizons. Permeability ranges between 10^{-3} and 10^{-5} m/s.

Cozzo dell'Aquila-Cozzo della Croce: the water body consists of deposits of the Terravecchia Unit. They consist of clays and marly clays, often with an abundant sandy fraction, ash-grey to bluish-grey in colour, with intercalations of large arenitic and conglomeratic beds. These rocks are discordant on the variegated clays, Numidian Flysch and grey marls of the Tortonian and are dated Upper Tortonian - Lower Messinian. The water body in its conglomeratic portions has an estimated primary porosity permeability of $10^{-3}-10^{-4}$ m/s. The permeability of the clay levels is almost zero (10^{-8} m/s). Groundwater circulation is discontinuous and confined within the conglomeratic levels, which, being between clay horizons, contribute to the definition of pressurised aquifers. The thicker conglomeratic lenses come into contact with the alluvium of the Belice Sinistro River and therefore have extensive hydraulic connections with it.

Rocca Busambra: the water body consists of a rocky succession referable to the Ibleo-Pelagiano domain (carbonate platform Saccense auct.).

The Rocca Busambra water body is characterised by rocky successions with some distinctive lithological features in the eastern and western sectors.

The western succession consists of, from bottom to top:

- Ivory white limestones with loferitic cyclothems of the Upper Trias - Middle Lias, approximately 400 m thick.
- Beige inland limestones of the Upper Lias - Dogger, a few metres thick.

In geometric conformity on, or disconformity on follow:

- Limestones and calcarenites, variable in colour, with ammonites, belemnites and aptians of the Dogger-Malm, for a thickness of a few tens of metres.
- Fine-grained limestones with flint, of the majolica or lattimusa type, greenish-white in colour, with haptens, brachiopods and calpionellae of the Upper Titonic - Lower Cretaceous, over a thickness of 10 to 20 m.
- Marly limestones of the scale type, globotruncane of the Upper Cretaceous, for several tens of metres.

The eastern succession consists, from bottom to top, of:

- Ivory white limestones with loferitic cyclothems of the Upper Trias - Middle Lias, for about 300-400 m in thickness.

Generally in geometric discontinuity, followed by:

- Breccias and conglomerates, with an almost completely absent matrix, which were formed by the dismantling of the Triassic-Leaphic platform, for a thickness of a few metres to about ten metres.
- Marly limestones and marly limestones, whitish or pinkish, globorotalia flake type, with thin calcarenite intercalations, dated to the Eocene and some ten metres thick.
- The sequence described indicates the transition from a platform environment to a predominantly pelagic one.
- The reliefs of Portella del Vento also belong to the body of water, formed from the bottom upwards:
- Ivory-white limestones, with loferitic cyclotomes, upwards intensely fractured, of the Upper Triassic - Middle Lias, approximately 50 m thick.

In conformity they follow:

- Limestones and reddish marly limestones of the Flake Type, with intercalations of calcarenites and graded brecciola, of the Upper Cretaceous, for a thickness of several tens of metres.
- Breccioli and conglomerates with a sparse matrix (if present, it is composed of marly limestone of the Flake Type), formed by the dismantling of the Triassic-Liassic platform, for a thickness of a few metres to about ten metres.
- Marly limestones, whitish or pinkish, globorotal Flake Type, with graded calcarenitic intercalations, dated Eocene and several tens of metres thick.
- The Pizzo Marabito relief also belongs to the Rocca Busambra body of water and shows, from bottom to top, a succession of:

- Ivory-white limestones, with loferitic cyclothems, generally fractured and brecciated, with a dense network of sedimentary strands consisting of reddish calcarenites to hinterland or vinaceous limestones to ammonites of variable age from Upper Lias to Dogger - Lower Malm.

In geometric disconformity, after an erosional gap, they follow:

- Very fine limestones and radiolarian limestones, vinaceous to greenish to yellowish in colour, almost completely silicified, from the Malm over a thickness of a few metres.
- Greenish limestones and limestone marls, lattimusa type, with haptens and calpionellas, of the Upper Malm - Lower Cretaceous for a thickness of tens of metres.
- Whitish or pinkish marly limestones and marly limestones, scale-type, with globotruncane and globorothal, of the Upper Cretaceous - Eocene, to a thickness of several tens of metres.

The entire fracturing present, with a sub-vertical trend, gives this complex a high permeability by cracking, demonstrated by the permeability coefficient that is greater than 10^{-2} m/s. The high areal development, together with its permeability characteristics, make this complex particularly interesting from a hydrogeological point of view. In fact, it can be considered both as one of the major absorption surfaces of precipitated water as well as one of the main preferential routes for the circulation of water in the area, which can find natural outlets in the springs (e.g. the Malvello and Drago springs, which originate from the contact between the Mesozoic limestones of Rocca Busambra and the San Cipirrello Marls).

It is assumed that there is a discontinuous hydraulic connection in the subsurface with the other carbonate bodies that outcrop to form the hydrostructures of the Rocche di Rao group.

9.2 Hydrogeology

The permeability of outcropping rocks can be categorised into:

- *Very high permeability through cracking.* The carbonate complex that forms the backbone of the entire Rocca Busambra and its adjacent reliefs belongs to this class. In fact, to this group belong the oolitic stromatolitic limestones of the Upper Trias–Lower Lias, the nodular limestones with crinoids belonging to the Red Ammonite (Dogger–Malm) and the carbonate megabreccia, sometimes dolomitised, of the Maastrichtian-Paleocene. These rocks have very little or no primary permeability. The intense, sub-vertical fracturing gives this complex a high permeability by cracking, with a permeability coefficient of more than 10^{-2} m/s. The high areal development in the territory, together with its permeability characteristics, make this complex particularly interesting from a hydrogeological point of view. It can be considered as one of the major surfaces for the absorption of precipitated water and one of the main preferential routes for the circulation of water in the area, which can find natural outlets in the springs (an example is the Malvello spring and the Drago spring, which flow from the contact between the Mesozoic limestone of Rocca Busambra and the Marne of San Cipirrello).

- *Medium to high permeability through cracking.* This class includes microcrystalline calcilutites, with planktonic organisms, immersed in a micritic matrix, alternating with marls, belonging to the Scaglia Auct. (Upper Cretaceous–Oligocene) and the white, porcelainized marly limestones with concoid fractures of the Lattimusa Unit (Upper Titonic–Neocomian) and the Hybla Unit (Middle Cretaceous). The presence of fractures within this hydrogeological complex gives it a secondary permeability; however, on the whole, the permeability coefficient (K) attributed to these soils has a value between 10^{-7} and 10^{-4} m/s. From these soils, in contact with the predominantly clayey complex, spurt the Fico, Marosa, Cerasa and Barone springs.
- *Very high permeability due to porosity.* Terrigenous deposits belonging to the Terravecchia Unit (Upper Tortonian–Lower Messinian) belong to this class. These sediments, which are scarcely permeable in the clayey portions, present discrete porosity permeability in correspondence with the sandy or conglomeratic levels. Overall, the value of the permeability coefficient (K) associated with this complex varies between 10^{-3} and 10^{-2} m/s.
- *Medium–high permeability due to porosity and cracking.* The quartzarenites of the Numidian Flysch (Upper Oligocene–Lower Miocene) belong to this complex. These deposits develop mainly in large banks with somewhat variable grain sizes, from medium-fine in the lower portion to medium-coarse in the upper portion. In general, permeability is determined not so much by porosity as by the degree of fracturing. The value of the permeability coefficient associated with this complex ranges from 10^{-4} to 10^{-2} m/s.
- *Very low permeability due to porosity.* Grounds belonging to the Mufara Unit (Carnic), the clays of the Numidian Flysch (Upper Oligocene–Lower Miocene) and the Marls of San Cipirrello (Langhiano–Lower Tortonian) can generally be grouped together within it, even though they may sometimes present some small characteristics that differentiate them. Overall, the permeability is very low, with a permeability coefficient of less than 10^{-9} m/s.

9.3 Hydrogeochemical Characterisation

In the Rocca Busambra hydrogeological basin three water bodies have been recognised. For the definition of the salient chemical-physical characteristics of this hydrogeological basin, four springs belonging to two distinct water bodies were considered (Fig. 9.4): one of a quartzarenitic nature relating to the Flysch (Mezzojuso water body, 1 sample) and one purely carbonate (Rocca Busambra waterbody, 3 samples). Sampled waters have temperatures ranging from 11.5 to 22.5 °C, which tend to progressively increase upstream to downstream, generally reflecting the trend of average seepage water temperatures. The pH shows a wide range

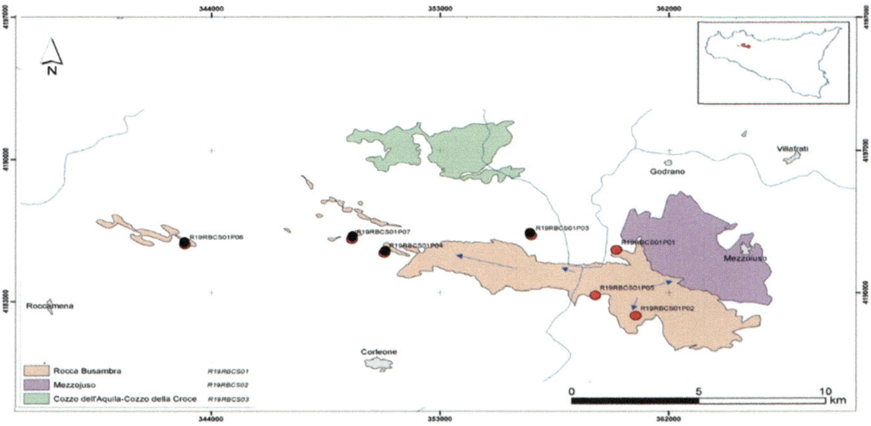

Fig. 9.4 Springs for drinking water use in the Rocca Busambra basin

between 6.5 and 8.3. Electrical conductivity values range between 219 and 695 μS/cm indicating a low degree of mineralisation of the water (Table 9.1).

In the Langelier-Ludwig diagram (Fig. 9.5), all spring waters plot in the bicarbonate-alkaline-earthy water quadrant.

The waters circulating in the Rocca Busambra water body, consisting of carbonate facies, have a bicarbonate-alkaline-earthy connotation due to the dissolution processes of calcite and dolomite.

Waters representative of the Mezzojuso water body, circulating in Flysch deposits, although falling in the same quadrant as the previous ones, show a tendency towards the Cl–SO_4 alkaline waters quadrant and are characterised by a wide compositional variability that reflects different salt contents but with similar ionic ratios (Na/Cl, Na/SO_4 and (Ca+Mg)/HCO_3). The enrichment in Na, Cl and SO_4 of these waters is clearly linked to the interaction with the numidic flysch, the origin of Ca and HCO_3 may be linked either to the dissolution of the carbonates of the rock, or to the calcite that forms the cement of the quartz arenites from which they flow.

In the ternary diagrams (Fig. 9.6) all the samples are arranged near the vertex of calcium and bicarbonate, being the most abundant dissolved ionic species.

Starting from these vertices, two trends can be distinguished: the first towards the vertex of Mg, probably linked to the dissolution of Triassic dolomitic rocks of the Trapanese Unit, and the second with a progressive increase in Cl and SO_4, corresponding to an increase in alkaline elements (Na and K), linked to the mineralogical complexity of the Flyschoid deposits. For further details see Alaimo and Ferla (1975), Bellanca et al. (1990), Favara et al. (2000), and Fontana et al. (2009).

Table 9.1 Chemical and isotopic composition of the sampled points

	T	EC	pH	Eh	Na	K	Ca	Mg	Cl	NO_3	SO_4	HCO_3	δ^2H	$\delta^{18}O$	WB
Brigna	12.5	673	8.33	197	1.99	0.083	4.47	1.54	2.06	nd	1.61	4.39	−43	−7.73	MJ
Drago	18.4	469	8.11	178	1.13	0.045	3.63	1.27	0.91	0.141	0.88	4.10	−42	−7.09	RB
Marosa	13.0	312	7.54	230	0.56	0.045	3.39	0.35	0.67	0.051	0.2	3.30	−43	−7.91	RB
Malvello	22.5	433	7.56	176	0.69	0.06	2.35	2.32	0.52	0.025	0.53	4.30	−44	−7.75	RB

Analytical data are expressed as meq/L, EC in μS/cm and δ^2H and $\delta^{18}O$ in ‰ versus SMOW. WB is for the waterbody. MJ is for Mezzojuso, and RB is for Rocca Busambra water bodies

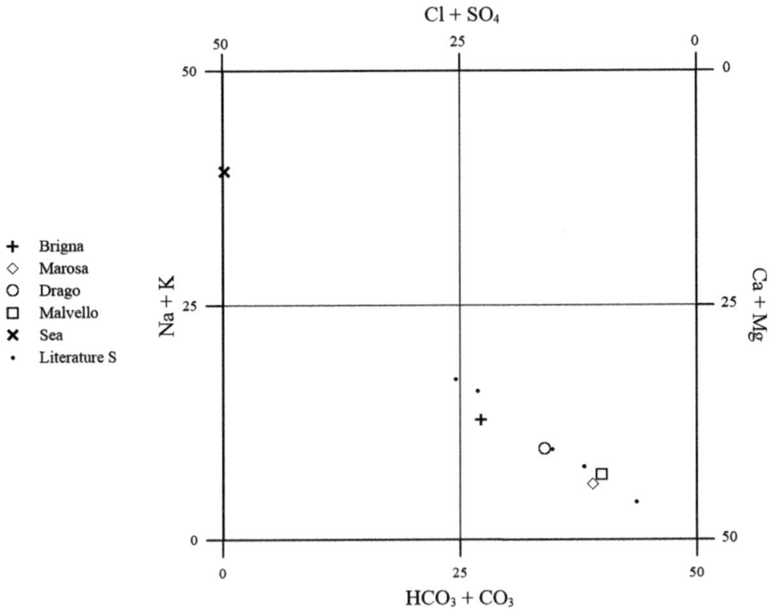

Fig. 9.5 Langelier–Ludwig classification diagram. Springs fall in the bicarbonate-alkaline-earthy. Literature data (Morici et al. 2020) are shown for comparison; S: spring, W: well

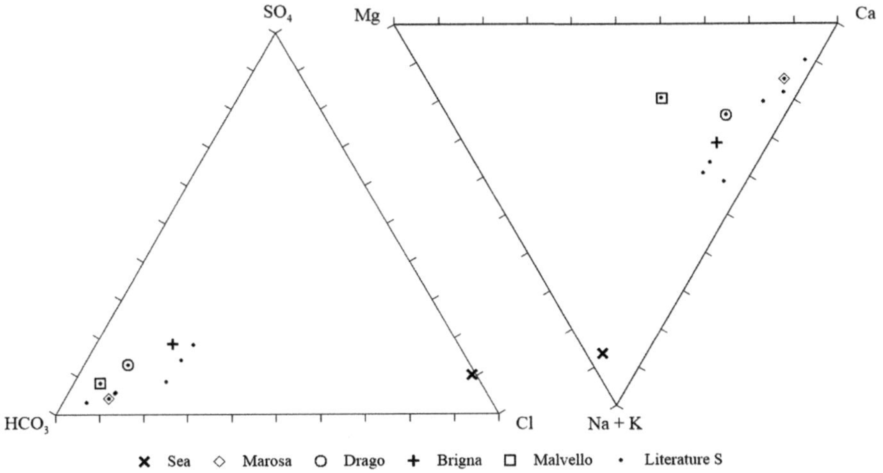

Fig. 9.6 Anion (left) and cation (right) ternary diagrams for Rocca Busambra catchment area. Literature data are shown (INGV 2005). S: Spring

9.4 Isotopic Characteristics of the Catchment Area

The isotopic composition of sampled water is characterised by δ^2H ranging from −44 and −42 δ‰ and δ^{18}O from −8 and −7 δ‰ versus SMOW (Table 9.1, this work, Morici et al. 2020). They fall between the MMWL and GMWL laying on the LMWL lines. The LMWLs are from Liotta et al. (2013); $\delta^{18}O = 6.31 * \delta^2H + 5$ as sector 1 (see Chap. 11 Table 1.6) and Fontana et al. (2009): $\delta^{18}O = 6.4 * \delta^2H + 5.5$ (Fig. 9.7).

Collected and literature water data fall close to the LMWL, confirming a predominant meteoric origin of these waters and the absence of secondary processes (like as drop evaporation).

The aquifers of the different water bodies appear to be predominantly fed by higher altitude waters. Considering an average isotope gradient of 0.27‰/100 m estimated by Fontana et al. (2009) specifically for the Rocca Busambra area ($\delta^{18}O = 0.00259{*}Q - 5.13$), it is possible to assume that all water bodies are fed from altitudes between 800 and 1.100 m (Fig. 9.8).

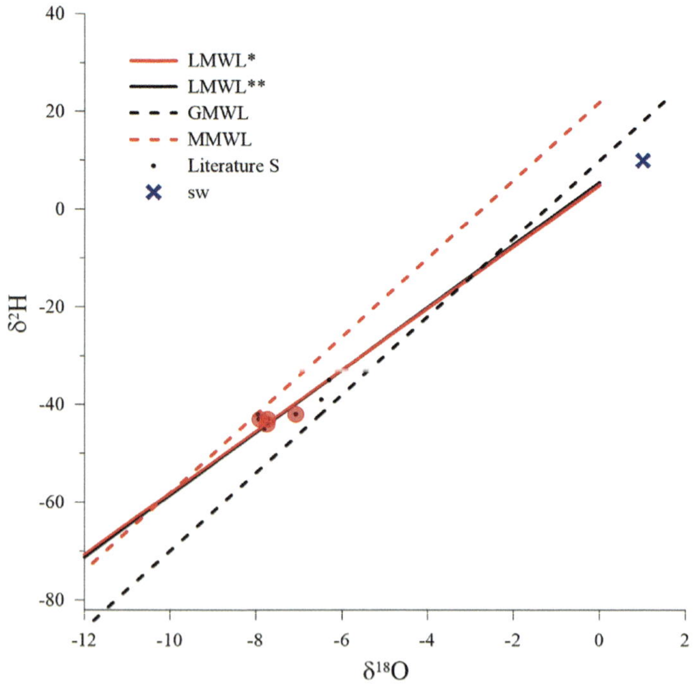

Fig. 9.7 δ^2H–δ^{18}O (in ‰ respect to SMOW) diagram of spring waters. The LMWL* is from Liotta et al. (2013), the LMWL** is from Fontana et al. (2009). The GMWL and MMWL are shown for comparison. Literature data from INGV (2005). S: Spring

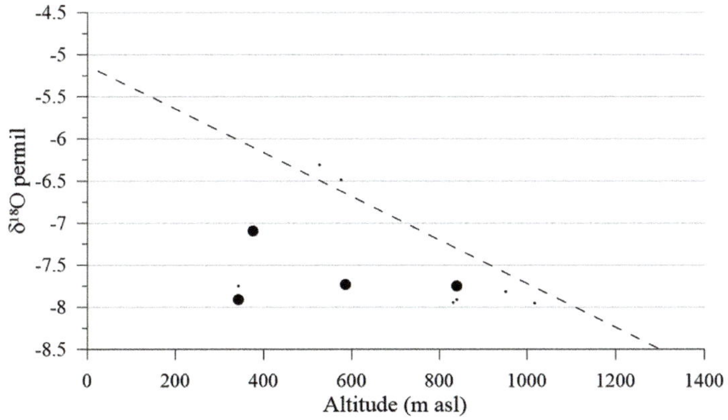

Fig. 9.8 Relationship between $\delta^{18}O$ and altitude

According to Fontana et al. (2009) two recharge zones exist, located (i) between 500 and 800 m a.s.l. corresponding with an outcrop of quartzarenites of the Numidian Flysch, and (ii) the highest recharge area (> 1.000 m a.s.l.) corresponding with the Rocca Busambra carbonate massif.

9.5 Trace Elements

Trace element concentrations in water springs (INGV 2005) are always below the legal limits for drinking water (Fig. 9.9, Table 1.1). Sr is the most concentrated element in all samples due to the carbonate nature of rocks, followed by Ba and B (Fig. 9.9; Table 9.2).

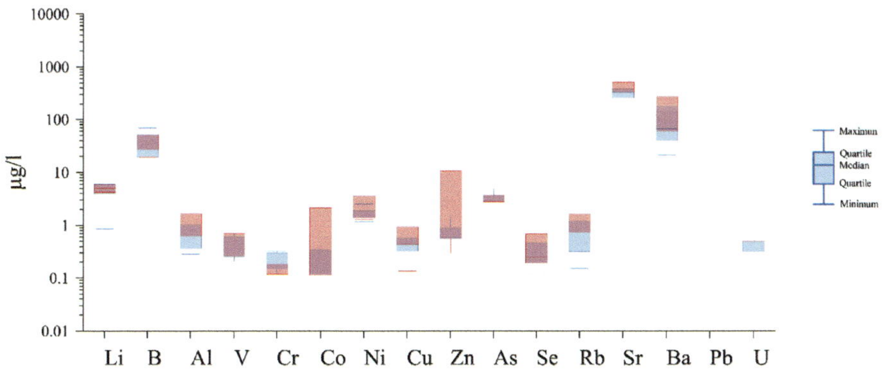

Fig. 9.9 Box-whisker plot of trace elements at Rocca Busambra basin. Red box spring data (this study). Light blue box data from INGV (2005)

Table 9.2 Trace elements in sampled spring waters

	Li	B	Al	V	Cr	Mn	Fe	Co	Ni	Cu	Zn	As	Se	Rb	Sr	Ba	Pb	U	WB
Brigna	4.24	19.6	1.680	0.25	bdl	760	21.0	2.14	3.65	0.58	10.7	0.62	0.69	1.67	136	34.6	0.073	bdl	MJ
Drago	6.00	52.3	1.040	0.61	0.12	7.75	9.40	0.11	1.48	0.44	0.57	3.77	0.3	0.736	329	60.2	bdl	0.494	RB
Marosa	3.84	27.3	0.578	0.71	0.18	bdl	7.70	bdl	1.30	0.94	0.30	2.73	0.19	0.36	513	278	bdl	bdl	RB
Malvello	5.97	36.0	0.632	0.26	bdl	bdl	3.00	bdl	1.38	0.14	0.67	2.83	0.13	1.22	440	65.5	bdl	0.506	RB

Concentrations are in μg/L

Mn reaches a very high concentration only in one sample (Brigna) in the Mezzojuso water body (Table 9.2).

Bibliography and Recommended Further Reading

Agate M, Basilone L, Catalano R, Franchino A, Merlini S, Sulli A (1998) Hypotheses on the structural condition of Rocca Busambra. In: Proceedings of the 79th National Congress Italian Geological Society, Palermo

Alaimo R, Ferla P (1975) Natrojarosite and thenardite, Narichsulphate hydrothermal minerals embedded within the dickite clay of Scillato-Caltavututro (Sicily). Period Min 44:227–243

Aniello S (2000) The agrarian landscape between enhancement and redevelopment at the boundaries of a 'protected area: use plan of the preriserva Bosco della Ficuzza, Rocca Busambra, Bosco del Cappelliere, Gorgo del Drago and its connection with local planning. Unpublished thesis, University of Palermo, Faculty of Architecture

Basilone L (2009) Mesozoic tectono-sedimentary evolution of Rocca Busambra in western Sicily. Facies 55:115–135. https://doi.org/10.1007/s10347-008-0156-2

Bellanca A, Dongarrà G, Neri R, Parello F (1990) Geochemistry of ground waters in carbonate rocks hosting fluorite mineralizations, northwestern Sicily. Acque Sotterranee IV:47–53

Catalano R, Lo Cicero G (eds) (1998) Sicily, a natural laboratory in the Mediterranean. Strutture, mari, risorse e rischi. 79th Congress Italian Geological Society, Palermo 18–20 september1998, Hikes guide, vol 1. Western Sicily, pp 185

Catalano R, Franchino A, Merlini S, Sulli A, Agate M, Basilone L (1998) Materials for Understanding the Deep Layout of West-Central Sicily. In: Catalano R, Lo Cicero G (eds) Western Sicily, Field Trip Guide of the 79° Congr. Naz. S. G. I., Mondello (Palermo), pp 175–185

Catalano R, Franchino A, Merlini S, Sulli A (2000) Central western Sicily structural setting interpreted from seismic reflection profiles. Mem Soc Geol It 55:5–16

Craig H (1961) Isotopic variations in meteoric waters. Science 133:1702–1703

Dall'Aglio M, Tedesco C (1968) Geochemical and hydrogeochemical study of springs in Sicily. Riv. Min. Sic.- n. 112–114, july– december 1968, 171–210

Favara R, Grassa F, Valenza M (2000) Hydrochemical evolution and environmental features of Salso River catchment, central Sicily (Italy). Environ Geol 39:1205–1215

Fontana M, Grassa F, Cusimano G, Favara R, Hauser S, Scaletta C (2009) Geochemistry and potential use of groundwater in the Rocca Busambra area (Sicily, Italy). Environ Geol 57:885–898

Gat JR, Carmi I (1970) Evolution of isotopic of atmospheric waters in the Mediterranean Sea area. J Geophis Res 75:1437–1440

Giannotti G, Lombardi L, Sidoti G (1972) Hydrogeological Chart of Western Sicily. Riv Min Sic n 133–135

Giunta G, Liguori V (1975) Considerations on the environmental significance and paleotectonic role of Rocca Busambra (Sicily). Boll Soc Nat 84:45–59, Napoli

Istituto Nazionale di Geofisica e Vulcanologia (INGV), Sezione di Palermo (2005) First characterization and monitoring of groundwater aimed at the drafting of the "Water Protection Plan of the Sicily Region". On behalf of the Sicilian Region – Deputy Commissioner for the waste emergency and water protection

Liguori V (1983) Integrated land study model (Ficuzza-Palermo). Technical geology and geomorphology. Agron Notebooks 10:33–52

Liotta M, Grassa F, D'Alessandro W, Favara R, Gagliano Candela E, Pisciotta A, Scaletta C (2013) Isotopic composition of precipitation and groundwater in Sicily, Italy. Appl Geoch 34:199–206

Longhitano S, Montanari L, Punturo R (1995) Genetic issues on Neptunian strands from Rocca Busambra (western Sicily). Boll Acc Gioenia Sci Nat 28:113–145

Morici S, La Pica L, Favara R, Gagliano Candela E, Pecoraino G, Scaletta C (2020) Chemical-physical parameters, major, minor and trace elements composition of groundwater belonging

to the hydrogeological basin Rocca Busambra in Sicily (Italy). Interdisciplinary Earth Data Alliance (IEDA). https://doi.org/10.26022/IEDA/111747
Of Public Works M, Hydrographic CSS (1934) The Italian springs – list and description. Publication No. 14 of the Service, vol II, Sicily. Istituto Poligrafico dello Stato, Rome
Ruggieri G (1966) Notes on the Miocene of western Sicily. Riv Min Sic 17(97–99):18–24
Wendt J (1959) Some new observations related to the stratigraphy of the Giurese in the Busambra Rock. (Palermo). Riv Min Sic 60:255–258

Sicani Mountains Hydrogeological Basin

10

Giovannella Pecoraino, Leonardo La Pica, Esterina Gagliano Candela, and Sabina Morici

10.1 Geological Framework

The Sicani Mountains are a vast area straddling the provinces of Agrigento and Palermo, largely made up of a hilly strip interspersed with vast plains occupied by arable and uncultivated land and pastures, while the mountainous area actual (over 800 m a.s.l.) is characterised by sheer cliff, mostly made up of carbonates (Fig. 10.1).

Six natural reserves are included in the Sicani area: Mt Carcaci, Mt Cammarata, Palazzo Adriano e Valle del Sosio Mts, Mt Genuardo and S. Maria del Bosco, Caves of Sant'Angelo. On the boundaries, other important natural reserves, such as Serre di Ciminna, Valle dello Iato and Mt San Calogero, create one of Sicily's most important ecological networks, in intimate connection with the Madonie Park and the reserves on the coastal strip.

The geological substratum of the Sicani Mountains is made up of Mesozoic carbonate rocks, generally massive or thickly stratified limestone; marl, dolomite, dolomitised or silicified limestone can also be found. Alterability is scarce or very scarce, porosity is very low, and permeability is usually high due to cracking, or very high due to karstification. The presence of fractures, faults and stratigraphic discontinuities has a decisive influence on the hydrographic network.

G. Pecoraino (✉) · L. La Pica · E. Gagliano Candela (Deceased) · S. Morici
Istituto Nazionale di Geofisica e Vulcanologia, Sezione di Palermo, Via Ugo La Malfa 153, 90146 Palermo, Italy
e-mail: giovannella.pecoraino@ingv.it

L. La Pica
e-mail: leonardo.lapica@ingv.it

S. Morici
e-mail: sabina.morici@ingv.it

E. Gagliano Candela et al. (eds.), *Drinking Water in Sicily*,
https://doi.org/10.1007/978-3-031-60473-7_10

Fig. 10.1 The Sicani Mountains. Imagine from Google Earth Pro v 7.3.6.9345 (December 29, 2022). Sicily, Italy

Figure 10.2 shows the simplified geological-structural scheme of western Sicily and Fig. 10.3 a structural scheme of the Sicanian Mountains. The pre-orogenic palaeogeography of this sector of the chain, proceeding from the internal to the external areas, can be schematised as follows:

- Imerese-Sicano Domain, an area of pelagic sedimentation that developed between two areas of predominantly neritic sedimentation, set on continental crust that has been thinning since the Triassic;
- Ibleo-Pelagiano Domain, a deep-sea area with complex physiography and condensed neritic and pelagic sedimentation in places, overall set on 'normal' thickness continental crust (Winnock 1981; Boccaletti et al. 1986, 1990; Ben-Avraham et al. 1990; Torelli et al. 1991).
- Imerese-Sicano Domain: represents an area of pelagic sedimentation between the Panormide Domain and the outer Ibleo-Pelagian Domain.

The successions of this wide pelagic sedimentation domain begin with siliceous-clastic rocks of the Lower-Middle Trias (Lercara Unit; Schmidt di Friedberg 1964), within which basic volcanites and breccias are found interspersed with elements of the Palaeozoic carbonate platform. This is followed by a succession consisting of clays, marls and calcilutites with pelagic lamellibranchs and frequent resedimented calcarenites (Mufara Unit; Schmidt di Friedberg et al. 1960) of Upper Trias age.

Within these two successions, basic volcanites are found interspersed at various levels. This sequence is followed by thin layers of calcilutites and flint nodules, radiolarians and pelagic lamellibranchs (Scillato Unit; Schmidt di Friedberg et al. 1960) of Upper Trias age.

The Lias successions consist of thick accumulations of resedimented dolomitized breccias and doloarenites (Fanusi Unit; Schmidt di Friedberg et al. 1960).

The Middle-Upper Jurassic successions are represented in the inner margin by radioliths, argillites and flinty marls, with frequent and voluminous additions of breccias and calcarenites, sometimes silicified, to carbonate platform elements (Crisanti Unit; Schmidt di Friedberg et al. 1960).

The Upper Cretaceous—Lower Oligocene successions consist of marly calcilutites and calcisylites with planktonic foraminifera of the 'Scaglia' Auct. type. These terrains are distributed throughout the basin area and are characterised by intercalations of carbonate breccias and megabreccias both to the north (Caltavuturo Unit; Schmidt di Friedberg et al. 1960) and to the south (Amerillo Unit; Patacca et al. 1979; Sestini and Flores 1986). Sedimentation ends with an alternation of marls, calcarenites and calcirudites with macroforaminifera of Oligocene age.

– *Ibleo-Pelagiano Domain*: represents the outermost palaeogeographic domain known in the outcrop. It is an area of neritic sedimentation since the Upper Triassic period, which, throughout the Mesozoic and Lower Tertiary, takes on the characteristics of a pelagic seabed.

The oldest successions consist of dolomitic, stromatolitic and loferitic limestones and limestones with megalodonts of the Upper Trias—Lias age (Inici Unit; Schmidt di Friedberg 1964). In the Upper Lias—Malm the platform successions are represented by condensed levels and nodular limestones with ammonites, crinoid calcarenites, radiolarian calcilutites and silicified foraminifers ("Rosso Ammonitico" and Buccheri Unit; Patacca et al. 1979). They outcrop in the sectors of Mt S. Calogero, Mt Magaggiaro, in the Pizzo Telegrafo mountain range, and at Mt Genuardo. This is followed by pelagic calcilutites with tintinnids and marls with haptens and planktonic foraminifera of the Upper Jurassic—Lower Cretaceous (Chiaramonte Unit and Hybla Unit; Patacca et al. 1979). The thickness of these rocks ranges from a few metres in the southernmost sectors (Mt Maranfusa, Pizzo Telegrafo, Mt S. Calogero di Sciacca) to a hundred metres in the areas that connect with the Imerese-Sicano Basin (Mt Kumeta, Mt Bonifato). This is followed by calcilutites and marls with planktonic foraminifera (Amerillo Unit; Patacca et al. 1979; eq. "Scaglia" Auct.) of the Upper Cretaceous-Eocene. In the Monte Genuardo-Pizzo Telegrafo areas, these deposits are interspersed with thick accumulations of carbonate breccias and carbonate platform elements. The Oligocene—Lower Miocene successions consist of calcarenites and calcirudites with macroforaminifers and calcareous algae, heterotopic to reef deposits in the southernmost areas (Mascle 1979; Pedley and Renda 1998).

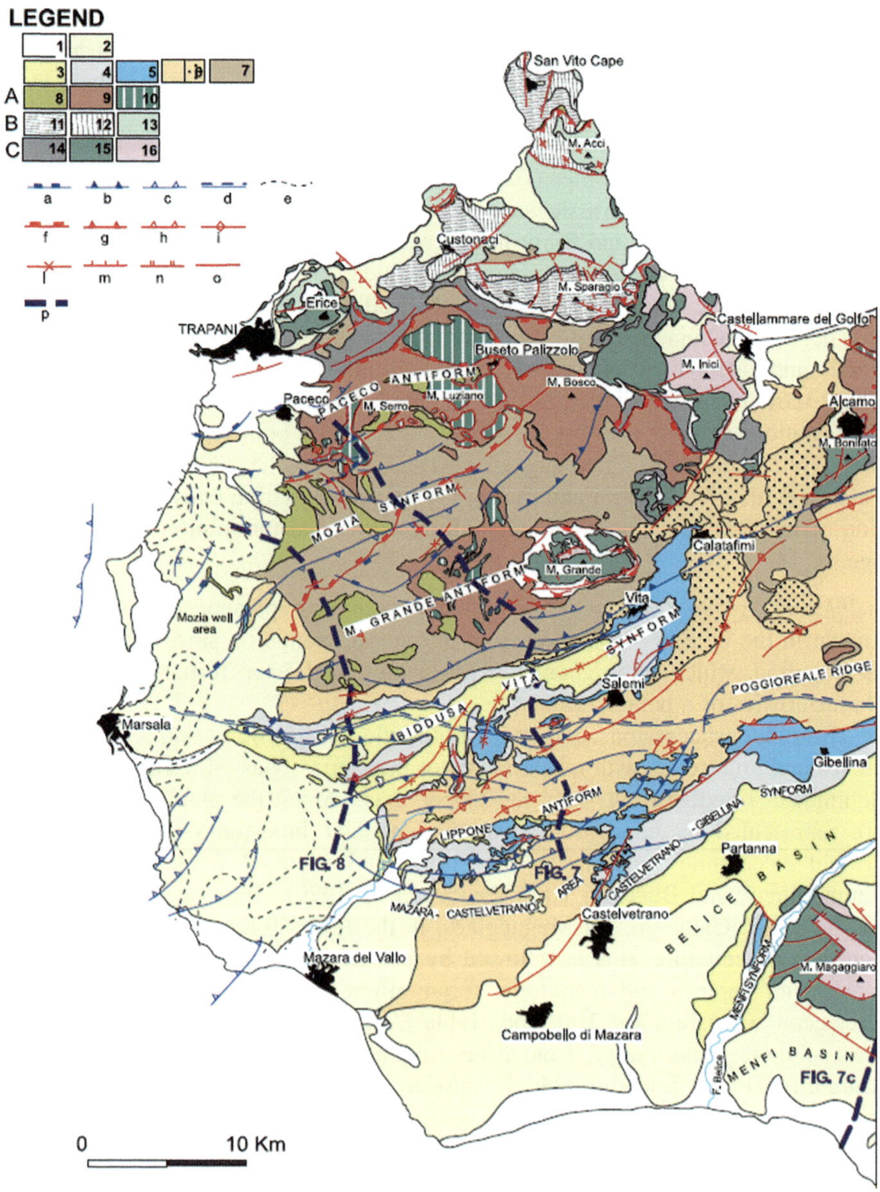
LEGEND
1 2
3 4 5 6 7
A 8 9 10
B 11 12 13
C 14 15 16
a b c d e
f g h i
l m n o
p
San Vito Cape
M. Acci
Custonaci
M. Sparagio
Erice
TRAPANI
Castellammare del Golfo
Buseto Palizzolo
M. Inici
PACECO ANTIFORM
M. Luziano
M. Bosco
Paceco
M. Serro
Alcamo
M. Bonifato
MOZIA SYNFORM
Calatafimi
M. Grande
M. GRANDE ANTIFORM
Mozia well area
Vita
VITA SYNFORM
Salemi
POGGIOREALE RIDGE
Marsala
BIDDUSA
Gibellina
GIBELLINA SYNFORM
LIPPONE ANTIFORM
CASTELVETRANO - GIBELLINA
Partanna
BELICE BASIN
FIG. 8
FIG. 7
MAZARA - CASTELVETRANO AREA
Castelvetrano
MENFI SYNFORM
M. Magaggiaro
Mazara del Vallo
Campobello di Mazara
F. Belice
MENFI BASIN
FIG. 7c
0 10 Km

◀**Fig. 10.2** Geological map of Sicani Mts Hydrogeological basin (modified from Catalano and Lo Cicero 1998). Legend: (1) Recent; (2) clastic carbonates, Pleistocene–Late Pliocene; (3) marls and calcarenites, Pliocene; (4) marly calcilutites (Trubi), Early Pliocene; (5) evaporites, Messinian; (6) reef carbonates, Early Messinian; Terravecchia Un., Early Messinian–LateTortonian. (A) *pre-Panormide units*: (7) marly arenaceous member (Langhian–Early Tortonian); (8) biocalcarenites and arenaceous strata (Mischio), Early Miocene; (9) calcarenites, shales and glauconitic quarzarenites, Oligocene; (10) marls and pelagic calcilutites with biocalcarenites, Early Cretaceous–Early Oligocene. (B) *Panormide units*: (7) marly arenaceous member, Langhian–Early Tortonian, glauconitic Calcarenites, Early Miocene; (11) shelf to slope calcarenites and calcilutites, Early Cretaceous–Oligocene; (12) calcilutites and calcarenites, Jurassic–Early Cretaceous; (13) platform carbonates and dolomites, Trias–Lias. (C) *Trapanese units:* (14) marly arenaceous member, Langhian–Early Tortonian; glauconitic calcarenites, Early Miocene; (15) pelagic platform deposits, Jurassic–Oligocene; 16) platform carbonates and dolomites, Trias–Lias. (a), (b), (c) buried thrust fronts; (d) pre-Panormide nappe fronts; (e) isobaths of the Plio-Pleistocene deposits; (f), (g), (h), thrust fronts in outcrop; (i), (l) fold axes; (m) extensional faults; (n) reverse faults; (o) strike-slip faults

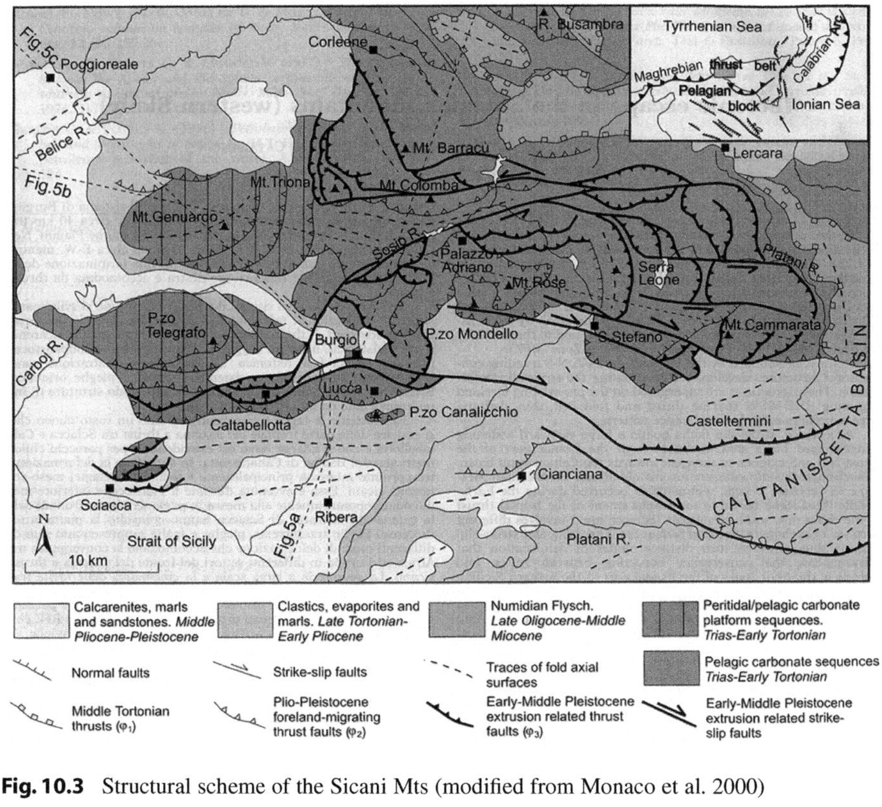

Fig. 10.3 Structural scheme of the Sicani Mts (modified from Monaco et al. 2000)

The syn-tectonic successions deposited in this sector of the Sicilian chain can be divided into various groups that embrace temporal intervals characterised by significant deformation phases:

Upper Oligocene—Middle Miocene Successions

The successions of this broad domain can be subdivided into numidic successions s.s. and successions within which there is a subordinate presence of numidic materials (Wezel 1973, 1974).

The representative numidic successions consist of:

- arenaceous facies, consisting of arenaceous deposits in layers and banks with pelitic intercalations. They outcrop along the alignment Monti di Trapani-Monti Iudica and Scalpello.
- pelitic-arenaceous facies, consisting of an alternation of pelites and sandstones in thin, rarely decimetric layers. The pelitic fraction is predominant compared to the arenaceous one.

They are stratigraphically superimposed on the facies described above and outcrop along approximately the same isopycnal belt.

Langhian—Middle Tortonian Successions

They consist of sandy pelites with planktonic foraminifera, rich in glauconite and pyrite; at various stratigraphic heights, there are intercalations of quartz sandstones and conglomeratic lenses, whose elements consist of carbonate fragments and numidic quartzarenites (S. Cipirello Unit; Ruggieri and Sprovieri 1970; and base of Licata Unit; Ogniben 1954). Deposition continues diachronically with the "Arenarie di Corleone" (Pedley and Renda 1998).

Upper Tortonian—Messinian Successions

The deposits consist of thick successions of sandy pelites and pelites with planktonic foraminifera, which change upwards to an alternation of sands and clayey sands, with intercalations of conglomerates, which become prevalent upwards (Unit Terravecchia; Flores 1959). In places, the normal sedimentary sequence is interrupted by strips of more ancient chaotic clays (A.B. I and A.B. II; Ogniben 1966), or carbonate or arenaceous olistolites, the latter of Numidian origin.

Upper Messinian Successions

The evaporitic successions can be grouped into two cycles (Decima and Wezel 1971), separated by an erosion surface and a regional amplitude discordance.

The successions consist of chalky clays, gypsumarenites (with intercalations of calcarenites in places), stromatolitic limestones passing through travertines, selenitic gypsums and laminated gypsums, the latter alternating with levels of clays of varying thickness. The succession is completed by microconglomerates, arcosic sandstones and micaceous silty marls ("Arenazzolo"; Decima and Wezel 1971), with subordinate levels of mudstones, gravels and sands.

Pliocene—Pleistocene successions

The base of the Lower Pliocene successions consists of marly and calcareous marly sediments with planktonic foraminifera (*'Trubi'*). These deposits rest indifferently on the Mesozoic-Tertiary substrate, separated by a regionally wide discordance surface. The *'Trubi'* consist mainly of an alternation of thinly laminated pelites and marly limestones, arranged in centimetric (or more rarely decimimetric) layers and with discrete lateral continuity. At various stratigraphic levels there are intercalations of clayey breccias and/or olistolites of various kinds (A.B. IV of Ogniben 1966), and they pass upwards, through a thick pelitic interval characterised by chaotic intervals (*slumps*), to blue marls with planktonic foraminifers with intercalations of clayey breccias (A.B. V of Ogniben 1966).

In the areas of Sciacca and the Belice Valley within the *'Trubi'* there are chaotic horizons due to syn-sedimentary landslides and intercalations of carbonate breccias (Mascle 1979). In these areas these deposits pass upwards to a succession consisting of an alternation of pelites, sandy pelites, graded and laminated quartz sandstones, with incorporated megabreccia from the underlying Pliocene substratum (Marnoso-Arenaceous Belice Unit; Ruggieri and Torre 1974). The succession is closed by a pack of biocalcarenites with thin intercalations of cross-stratified sandy clays, which upwards evolve into conglomerates, biocalcarenites and sands (Belice Calcarenites; Ruggieri and Unti 1977).

10.2 Hydrogeology

The water bodies in the Sicani Mountains have different hydraulic characteristics and potential, in relation to the degree and type of permeability of the rocky media and their geometric characteristics (hydrostructures).

Terrains with different permeability can be distinguished:

- Medium–high permeability due to porosity: this class includes the calcarenites, glauconitic calcarenites, oolitic calcarenites and Oligocene marls that outcrop in the eastern and central-western sectors of the area. From a hydrogeological point of view, although these soils have good permeability, they are not the site of particularly interesting aquifers.
- High permeability due to primary porosity: the colluvial deposits to the south of Mt Carcaci, to the south-east of Prizzi and to the south of Palazzo Adriano, the groundwater debris along the eastern slopes of Mt Genuardo, along the southern slopes of Mt Colomba and finally on the western slopes of Mt Cassaro belong to this class. No springs with a significant flow rate were found for the present study, except at Portella Saracena (south of Palazzo Adriano) where at an altitude of 631 m a.s.l. the San Benedetto spring is present with a flow rate of 1.3 l/s.

- Low permeability due to cracking: marly and calcareous marly soils (*'Trubi'*) belong to this class. These outcrops are largely present to the west and east of the village of Giuliana. No springs have been found in these terrains.
- Medium–high permeability due to cracking: marly limestones, calcilutites and argillites distributed variously over the study area. From a hydrogeological point of view, these soils are linked to springs with very limited flow rates (< 1 l/s), except to the east of Filaga, where the Zarchia spring is present with a flow rate of 8 l/s.
- High permeability due to cracking and/or karstification: to this category belong all the carbonate outcrops that are predominantly present in the area and to which the major springs are linked, such as those of Mt Scuro, Fontana Grande, etc.
- Practically impermeable: clayey and clayey-marly terrains in the eastern, central and western parts of the area belong to this class. These lithologies have almost zero permeability in clayey terms and very low permeability in marly and clayey-marly terms. Based on their position and their relationship with the rocks in which the water tables are located, they play a decisive hydrogeological role as a buffer.

The main hydrogeological complexes identified in the Sicani Mountain area are:

Calcareous-marly and clay-arenaceous-calcareous complex

This complex is made up of clays and variegated marls with intercalated calcilutitic, calcisiltitic and calcarenitic levels and thick lenses of dolomitic breccias (Mufara Auct Unit, Carnic). This complex, as a whole, has a very low permeability, which is further reduced when it is intensely tectonised. This complex represents the defined permeability limit of the overlying calcareous and calcareous-dolomitic complex (Mt Genuardo, Pizzo Telegrafo, Mt delle Rose and Mt Cammarata), with a relative impermeable character. The vulnerability is low and practically negligible.

Calcareous and calcareous-dolomitic complex

These rocks are calcilutites, calcarenites, dolomitic breccias, dolorudites and doloarenites pertaining to the Upper Triassic—Jurassic. Its hydrogeological characteristics render them one of the main aquifers of the Sicani Mountains. The complex is intensely fractured.

The prevailing permeability is high due to fracturing, karstification and, in places, porosity, when the complex is intensely fractured.

Vulnerability is high to very high, depending on the degree of karstification of the limestones or fracturing.

Calcareous and silico-marly complex
Grouped in this complex are sequences of siliceous argillites, radiolarian and radiolarian marls, nodular limestones, calcarenites and silicified calcisylites. These deposits of Jurassic—Cretaceous age. The permeability by cracking is low and increases locally in calcarenite successions. Overall, however, the permeability is on average low. Vulnerability is considered medium to low.

Calcareous-marly complex
This complex is predominantly composed of a calcareous-marly and marly succession with calcarenite intercalations. In the complex, the permeability through cracking is low, although it is higher at the intercalations of calcarenites and megabreccias.

Clayey-marly and marly-clayey complex
In this complex the Oligocene—Miocene and Pliocene clayey marls and clayey marls are grouped, having a low degree of permeability and therefore constituting the relative impermeability of the carbonate and arenaceous successions.

Calcarenitic-marly complex
This complex comprises the glauconitic calcarenites found in the northern sector of the Sicani Mountains. The permeability of the calcarenite deposits is sometimes high when it shows a dense network of fracturing.

Calcareous-chalky complex
It comprises the evaporitic deposits (evaporitic limestone, gypsum, gypsum arenites and gypsum clays) of the Messinian Gypsum-Sulphur Series. They outcrop in correspondence with Caltabellotta, Lucca and on the margins of the Sicani in correspondence with Casteltermini, Sant'Angelo Muxaro. Their permeability varies in relation to the degree of fracturing and karstification.

Calcarenitic and sandy complex
This complex includes calcarenite and sandy deposits from the Plio—Pleistocene. These terrains have a permeability that varies from high to medium–low depending on the grain size, the degree of cementation and fracturing and the presence of clay intercalations. It stretches in correspondence with Menfi-Capo S. Marco and Montevago.

The following water bodies have been distinguished in the Sicani Mountains (Fig. 10.3).

On the basis of the hydrogeological characteristics of the area, the following water bodies have been distinguished in the Sicani hydrological basin:

- **Menfi-Capo San Marco**: is composed of a Plio—Pleistocene rocky succession consisting of biocalcarenites with thin intercalations of cross-stratified sandy clays, evolving upwards to conglomerates, biocalcarenites and sands. Terraced deposits consisting of calcarenites, transitioning upwards to polygenic conglomerates, close the succession. The Pleistocene calcarenites lie on the clayey and clayey-marly deposits of the Middle-Upper Pliocene that represent the impermeable substratum. These hydrogeological conditions, associated with the presence of clayey intercalations at several levels, give rise to a multi-hydrogeological structure where the most consistent is certainly the basal one. The permeability of the rocky medium is medium–high due to primary porosity (10^{-3} cm/s).

- **Mount Magaggiaro**: the water body consists of a rocky succession referable to the Ibleo-Pelagiano domain (carbonate platform Saccense auct.). From bottom to top, the stratigraphic succession is represented by:
  - limestones and dolomitic limestones of the carbonate platform. Grey-white megalodontide limestones, often strongly dolomitised, with massive or indistinct stratification, passing upwards to white limestones, well stratified in large beds. These deposits are cyclically alternated with stromatolitic beds and, in places, with levels of greenish azoic marls (Lower-Middle Lias).
  - Ammonitiferous limestones: a succession of predominantly condensed, laterally discontinuous lithostratigraphic units that can be identified in outcrop on the basis of physical limits consisting of erosion or non-depositional surfaces. These deposits, characterised by rich associations with ammonites, haptites, belemnites and brachiopods, have an average thickness of 10–20 m. From the bottom upwards they differentiate: (i) hinterland calcarenites in discontinuous lenses (Pliensbachian—Toarcian); (ii) reddish-grey calcilutites, biocalcarenites and biocalcirudites with Posidonia Alpina, with indistinct to massive stratification and alternating centimetric levels of biocalcarenites with lagenides (Dogger); (iii) nodular, green to pink marly calcilutites with frequent encrusted lenses (Oxfordian—Kimmeridgian); (iv) white calcilutites with calpionella in 50–80 cm thick layers alternating with graded levels of biocalcarenites and biocalcirudites (Titonic);
  - marly calcilutites with rare, yellowish-white, sometimes pseudonodular flint strips and frequent slumping structures, containing tintinnids, radiolarians and sponge spicules. The total thickness is about 20 m. These deposits transition upwards to marly calcilutites and white marls with planktonic foraminifera, with abundant nodules and black flint lists, followed by greenish marls with haptens. The thickness is about 40 m; the age is Upper Titonic—Albian;
  - white marly calcilutites with nodules and lists of black or yellow flint (Scaglia) with planktonic foraminifera, transitioning upwards to white or pinkish calcilutites, well stratified in decimetric plan-parallel levels with rare intercalations of resedimented calcarenitic and calciruditic levels and frequent slumping structures. The age is Cenomanian—Eocene;
  - marly limestones and white marls with intercalated levels of resedimented biocalcarenites. The thickness is about 50 m, the age is Middle to Upper Oligocene;
  - marls and marly limestones: pelitic outer platform deposits composed of greenish and grey sandy marls more than 50 m thick, frequently alternating

with ochreous limestone-marly levels 0.5 m to 1 m thick. The age is Upper Oligocene;
- pinkish-grey limestones with glauconite granules, sometimes dolomitised, laterally transitioning to poorly cemented yellowish biocalcarenites with thicknesses ranging from 20 to 50 m. These deposits present a partitioning into flat-parallel banks of 50–100 cm, sometimes with cross lamination. The base of these deposits is marked by a discordant surface with sometimes pronounced erosional truncation. The age is Lower Miocene;

- **Monte Genuardo**: consists of a rocky succession referable to the transition zone between the Ibleo-Pelagiano domain (Saccense carbonate platform auct.) and the Imerese-Sicano domain (Sicano basin auct.). From bottom to top the stratigraphic succession is represented by:
  - massive, whitish, carbonate platform dolomites and dolomitic limestones consisting mainly of bioconstructed carbonates and reef breccias. There are stromatolitic and loferitic dolomites alternating with vacuolar megalodont dolomites. The thickness is about 80 m, the age is Noric—Rhaetic;
  - hazel-white resedimented bioclastic calcarenites in centimetric plan-parallel to curved non-parallel layers alternating with thin layers of pelagic to radiolarian calcilutites. Lamination, normal grading and plane-parallel grading are frequent. They reach a maximum thickness of 80 m. At the roof of the oolitic limestones there are lenses of reddish crinoid limestones of the Pliensbachian. The age is Lower-Middle Lias;
  - dark grey pelagic calcilutites, in places pinkish and nodular, in centimetric to decimetre layers, with rare interlayers of grey or greenish marls. The limestones consist of radiolarian biomicrites and sponge spicules. The thickness is about 80–100 m. At the base there are metric wedge-shaped intercalations of breccias, with inverted gradation and decimetric-centimetric elements, of Triassic reef carbonates. Lenticular bodies of basalts are found in the summit portion and at the transition with the overlying radiolarites. The age of the succession is Lower Toarcian—Oxfordian;
  - polychrome radiolarites and calcilutites with interbedded silicified radiolarian marls, in centimetric plan-parallel layers. The thickness is 20–50 m; the age is Oxfordian-Lower Titonic;
  - thinly stratified, white, haptic calcareous marls and marls, transitioning upwards to white calcilutites, with flint (Lattimusa) laths and nodules, with thin intercalations of greenish-grey marls. The calcilutites consist of calpion biomicrites. The thickness is about 50 m; the age is Titonic—Berriasian;
  - Aptian marls and flinty calcilutites (eq. Un. Hybla): green marls with rare centimetric intercalations of greenish flinty calcilutites consisting of Aptian wackestone. The thickness is about 50 m; the age is Aptian–Albian;
  - white, pinkish or red globotruncane calcilutites in Scaglia facies, generally discordant on Lower Cretaceous or Lattimusa deposits and radiolarites. The succession is approximately 50 m thick, whereas the age is Campanian;
  - megabreccia: ruditic-tabular or wedge-shaped bodies of carbonate breccias and megabreccias, in clinostratified places, discordant on the Cretaceous

Scaglia. The maximum thickness is about 100 m. The decimetric-centimetric elements, which are rarely metric, consist of Triassic—Liassic peritidal carbonates and more rarely of more or less silicified crinoid calcarenites, basalts and pelagic carbonates of the Lower Jurassic and Cretaceous. The matrix, which is rare, consists of Scaglia. The age is Maastrichtian;
  - white marly calcilutites (Scaglia) with flint lists and thin interlayers of greenish marls. In the upper portion there are decimetric levels of calcirudites and bioclastic calcarenites. The thickness is approximately 50 m; the age is Eocene;
  - greenish-grey or reddish sandy marls with planktonic foraminifera, with a maximum thickness of 60–70 m, transitioning upwards to cross-stratified quartzitic-calcareous-glauconitic arenites. Laterally and upwards there are grey sandy marls 100 m to 200 m thick. The age is Oligocene;
- **Montevago**: consists of yellowish-coloured, well-cemented and porous clastic and bioclastic deposits of Pleistocene age in which there are discontinuous clay and marly intercalations whose thickness does not exceed 2 m. The maximum thickness of the entire calcarenitic body varies between 45 and 75 m. The Pleistocene deposits lie stratigraphically on the clays and sandy clays of the Middle to Upper Pliocene, which are present to the west along the valley of the Belice River, and to the north and north-east in the middle to upper part of the left slope of the Senore stream. To the east and south, on the other hand, it is possible to find the Pleistocene calcarenites in direct contact with those of the Upper Pliocene; this contact, which is certainly discordant, can be observed approximately 2 km from Portella Misilbesi, in a north-westerly direction. In the southern sector, the Middle-Upper Pliocene calcarenites lie, in transgression, directly on the Mesozoic carbonate deposits of Monte Magaggiaro. The morphological layout of the territory, which is predominantly flat, masks the tectonics of the carbonate substratum, which abruptly subsides as one moves northwards.
- **Southern Saccense**: consists of a rocky succession referable to the connection zone between the Ibleo-Pelagiano domain (carbonate platform Saccense auct.) and the Imerese-Sicano domain (basin Sicano auct.). From bottom to top, the stratigraphic succession is represented by:
  - limestone and dolomitic limestone of carbonate platform. Grey-white megalodontide limestone, often strongly dolomitised, with massive or indistinct stratification, passing upwards to white limestone, well stratified in large beds. These deposits are cyclically alternated with stromatolite beds and, in places, with levels of greenish azoic marls (Lower-Middle Lias);
  - Ammonitiferous limestone: a succession of predominantly condensed, laterally discontinuous lithostratigraphic units that can be identified in outcrop on the basis of physical limits consisting of erosion or non-depositional surfaces. These deposits, characterised by rich associations with ammonites, haptites, belemnites and brachiopods, have an average thickness of 10–20 m. From the bottom upwards they differentiate: (i) hinterland calcarenites in discontinuous lenses (Pliensbachian—Toarcian); (ii) reddish-grey calcilutites,

biocalcarenites and biocalcirudites with Posidonia Alpina, with indistinct to massive stratification and alternating centimetric levels of biocalcarenites with lagenides (Dogger); (iii) nodular marly calcilutites, green to pink in colour, with frequent encrinitic lenses (Oxfordian—Kimmeridgian); (iv) white calcilutites in 50–80 cm thick layers of calpionella alternating with graded levels of biocalcarenites and biocalcirudites (Titonic);

- marly calcilutites with rare, yellowish-white, sometimes pseudonodular flint strips and frequent slumping structures, containing tintinnids, radiolarians and sponge spicules. The total thickness is about 20 m. These deposits transition upwards to marly calcilutites and white marls with planktonic foraminifera, with abundant nodules and black flint lists, followed by greenish marls with haptens. The thickness is about 40 m, while the age is Upper Titonic—Albian;
- megabreccia: yellowish-white chalcirudites in large beds with inconspicuous plan-parallel to clinoform layering, organised in tabular or wedge-shaped lithosomes. The elements are centimetric to metric in size and consist mostly of peritidal carbonates and subordinately of oolitic limestones, encrinites and basalts. The matrix, which is rare, consists of white globotruncane calcilutites. It is about 150 m thick and rests discordantly on the Liassic-Triassic carbonate platform soils, traversed by a dense network of fractures;
- white marly calcilutites with nodules and lists of black or yellow flint (Scaglia) with planktonic foraminifera, transitioning upwards to white or pinkish calcilutites, well stratified in decimetric plan-parallel levels with rare intercalations of resedimented calcarenitic and calciruditic levels and frequent slumping structures. The age is Cenomanian—Eocene;
- marly limestone and white marls with intercalated levels of resedimented biocalcarenites. The thickness is about 50 m; the age is Middle to Upper Oligocene;
- marls and marly limestones: pelitic outer platform deposits composed of greenish and grey sandy marls more than 50 m thick, frequently alternating with ochreous limestone-marly levels 0.5 m to 1 m thick; Upper Oligocene age;
- Pinkish-grey limestone with glauconite granules, sometimes dolomitised, laterally transitioning to poorly cemented yellowish biocalcarenites with thicknesses ranging from 20 to 50 m. These deposits present a partitioning into flat-parallel banks of 50–100 cm, sometimes with cross-laminations The base of these deposits is marked by a discordant surface with sometimes pronounced erosional truncation. The age is Lower Miocene;

- **Nothern Sicani**: consists of a rocky succession referable to the Imerese-Sicano domain (Sicano auct. basin), which consists almost entirely of the intensely dolomitised flint-bearing calcilutites of the Upper Trias. At the base, some patches of the Mufara Un. outcrop. The average permeability of the sub-hydrostructure is high due to cracking and karstification and is variable within the different portions of the succession.

- **Southern Sicani**: The sub-hydrostructure consists of a rock succession referable to the Imerese-Sicano domain (Sicano auct. basin). The base of the succession consists of the intensely dolomitised flint-bearing calcilutites of the Upper Trias. Towards the top, the succession shows marked similarities with that of the Monte Genuardo Unit and is characterised by resedimented oolitic calcarenites and ammonite calcilutites with pillow basalts of the Lower-Middle Jurassic followed by calcilutites and marls, sometimes siliceous, of the Upper Jura-Lower Cretaceous. This is followed in discordance by Upper Cretaceous megabreccias, 'Scaglia' deposits and Oligo-Miocene marls and sandstones.
- **Central Sicani**: consists of a rock succession referable to the Imerese-Sicano domain (Sicano auct. basin). The lithological succession consists of limestones with flint from the Upper Triassic followed by discontinuous patches of Jurassic deposits. In places, there are resedimented oolitic calcarenites of the Lias, followed by radiolarites and flint-bearing calcilutites of the Middle-Upper Jurassic. Jurassic deposits consisting mainly of flinty limestones and radiolarites are present in a discontinuous manner. This is followed by Cretaceous—Eocene deposits of the Scaglia containing, in places, levels of megabreccias. On the Scaglia there are Oligocene marls and, in discordance, glauconitic sandstones arguably of the Aquitanian—Langhiano. These are followed upwards by marls and sandy clays of the Serravallian—Tortonian. Basaltic volcanites (alkaline basalts) are present in these units and have been considered to be of Eocene age.
- **Eastern Sicani**: consists of a rock succession referable to the Imerese-Sicano domain (Sicano auct. basin). The rock succession is characterised by calcilutites with flint from the Upper Triassic passing upwards to marls and polychrome calcilutites and radiolarites from the Lower and Middle Jurassic with, in places, intercalations of calcarenites and resedimented carbonate breccias. On these deposits, rest the calpion limestones and marls of the Upper Jurassic—Lower Cretaceous. This is followed upwards by the Cretaceous—Eocene calcilutites of the 'Scaglia', with intercalations of nummulite breccias in the terminal portion and marls with planktonic foraminifera from the Oligocene with modest sandstone intercalations. Glauconitic biocalcarenites (Burdigalian—Langhiano) and clayey marls with Orbulina (Serravallian—Lower Tortonian) are found in discordance.

10.3 Hydrogeochemical Characterisation

Analyses of 23 springs representative of 9 water bodies were considered in this basin (Fig. 10.4) and listed in Table 10.1. Literature data (INGV 2005) are show for comparison. Sampled waters Showa wide range of values in chemical-physics parameters: temperature varies between 12 and 31 °C, EC between 340 and 2000 μ S/cm and pH between 6.8 and 8.3.

All the sampled spring waters plot within the quadrant representing bicarbonate-alkaline-earthy waters (Fig. 10.5). The cloud of samples close to the calcium and bicarbonate vertex (Fig. 10.6) is related to contact with carbonate rocks prevalent in the basin. A trend towards values with higher Na + K concentrations, evident in Figs. 10.5 and 10.6, is probably linked to ion exchange phenomena with marly-clay lithofacies.

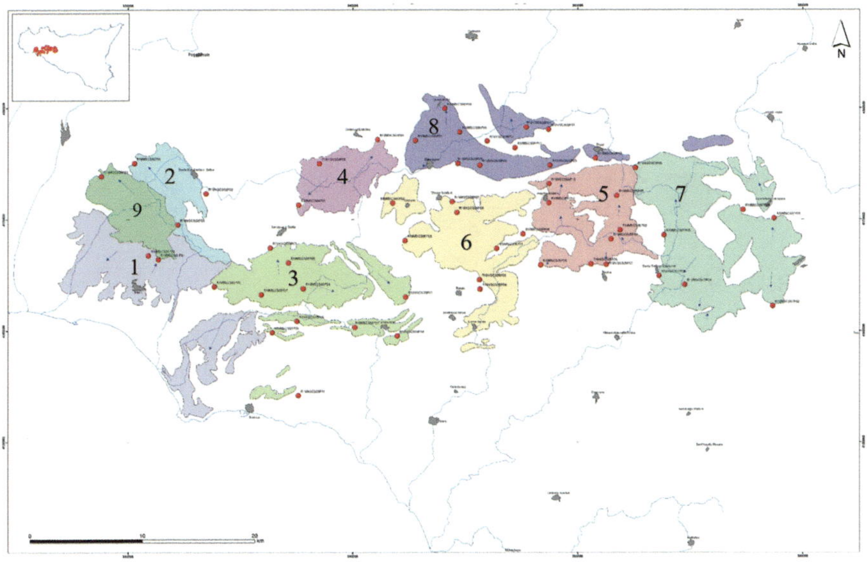

Fig. 10.4 Map of Sicani hydrological basin. Red dots literature points (INGV 2005), Black dots this study). Numbers identify water bodies: (1) Menfi-Capo S.Marco; (2) Montevago; (3) Southern Saccense; (4) Mt Genuardo; (5) Central Sicani; (6) Southern Sicani; (7) Easthern Sicani; (8) Northern Sicani; (9) Mt Magaggiaro. The location of springs is shown (green plots represent data from PRGA and INGV (2005), blue dots spring sampled for this work)

Table 10.1 Chemical and isotopic composition of the sampled waters

	T	EC	pH	Eh	Na	K	Ca	Mg	Cl	NO_3	SO_4	HCO_3	$\delta^2 H$	$\delta^{18}O$	WB
Alvano	14.2	548	7.2	148	0.72	0.04	5.71	0.32	0.7	0.374	0.34	5.45	−37	−6.55	MG
Battellaro	13.9	534	7.31	138	0.66	0.038	5.59	0.28	0.56	0.295	0.44	5.7	−37	−6.86	MG
Salto	18.4	618	7.22	218	1.8	0.113	4.54	0.83	1.74	0.239	0.6	4.5	−32	−5.97	ScM
Catafaldi	18.5	508	7.5	174	0.85	0.056	3.71	1.72	1.3	0.122	0.44	4.35	−33	−6.21	ScM
Figarella	16.3	832	7.12	219	1.43	0.0674	7.5	1.36	1.67	0.777	1.41	6.5	−34	−6.19	ScM
S. Matteo	14.0	355	8.04	144	0.33	0.023	2.41	1.35	0.47	0.017	0.55	3.00	−42	−7.64	SC
Galline	14.6	558	7.38	271	0.71	0.032	4.97	1.28	1.3	0.042	0.37	5.3	−39	−7.02	SM
Fico Bassa	15.6	693	7.08	204	0.98	0.082	7.26	0.47	0.94	0.453	1.26	5.95	−37	−6.76	SM
Chirumbo	13.2	365	7.51	98	0.43	0.031	2.4	1.51	0.36	0.008	1.21	2.86	−40	−7.5	SO
Prisa	12.8	516	7.33	188	0.57	0.055	4.88	1.00	0.46	0.211	1.08	4.8	−39	−7.08	SO
Martinazzo	14.2	546	7.08	194	0.58	0.022	5.5	0.45	0.84	0.2	1.09	4.4	−38	−6.9	SS
Scorciavacche	15.4	357	7.68	216	0.44	0.017	3.79	0.23	0.57	0.317	0.18	3.4	−39	−6.83	SS
Raia	16.0	506	7.75	531	0.83	0.026	5.00	0.39	0.67	0.05	0.46	5.15	−41	−7.29	SS
Madonna delle Scale	14.3	474	7.64	213	0.61	0.032	3.88	1.17	0.62	0.037	0.64	4.5	−39	−7.09	SS
Gragotta Grande	14.1		6.85		0.44	0.026	2.84	1.73	0.51	0.081	0.52	4.2			
Gragotta Piccola	14.3		7.35		0.35	0.026	2.79	1.89	0.59	0.081	0.87	4.03			
Fico Granatelli	15.9		6.5		1.05	0.051	3.49	1.4	0.68	0.097	1.27	3.62			
Innamorata	14.7		6.8		0.31	0.2	2.69	1.65	0.68	0.032	0.17	4.13			
Casale	12.6		6.85		0.48	0.02	2.99	0.99	0.59	0.065	0.19	3.79	−40	−7.28	SM
Liste di Sciacca	16.4		6.95		0.48	0.051	2.74	1.81	0.56	0.016	0.63	3.95	−41	−7.68	SO

(continued)

Table 10.1 (continued)

	T	EC	pH	Eh	Na	K	Ca	Mg	Cl	NO_3	SO_4	HCO_3	δ^2 H	$\delta^{18}O$	WB
Fontana grande	12.8		7.1		0.26	0.01	2.2	1.73	0.51	0.048	0.13	3.46	−42	−7.61	SC
Capo D' Acqua	13.4	339	7.28	206	0.36	0.031	2.69	1.61	0.38	0.053	0.41	3.9	−41	−7.75	SC
S. Rosalia	14.5	405	7.65	139	0.38	0.033	3.01	1.68	0.37	0.08	0.43	4.3	−40	−7.2	SC

Analytical data are expressed as meq/L, EC in μ S/cm and δ^2H and δ^{18}O in ‰ versus SMOW. WB is for water body. MG is for Mt Genuardo, ScM is Southern Saccense, SC is Central Sicani, SM is Southern Sicani, Sw is Weasthern Sicani, SS is for Northern Sicani water bodies

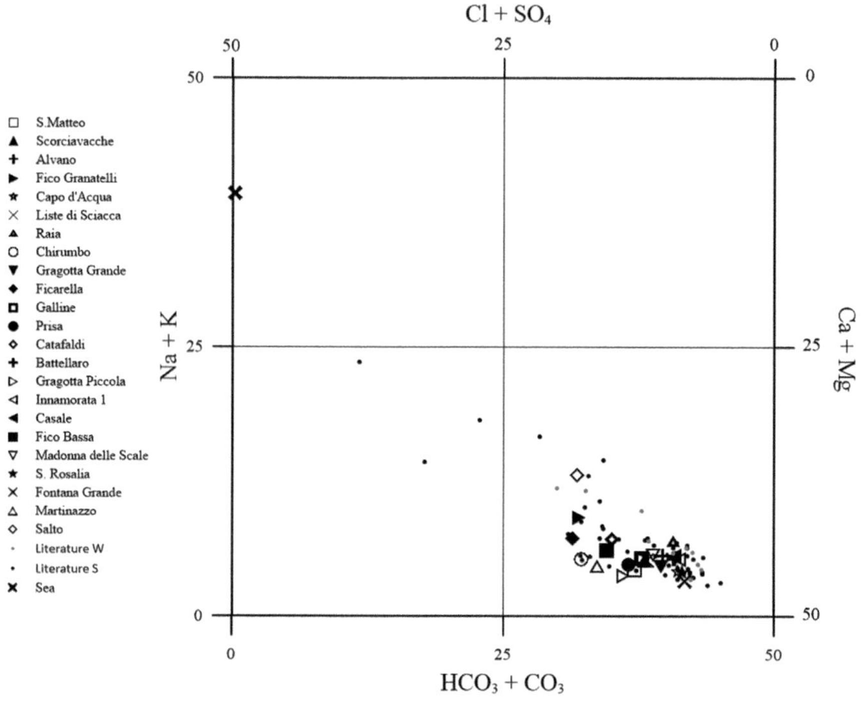

Fig. 10.5 Langelier–Ludwig classification diagram. The springs fall in the bicarbonate-alkaline-earthy quadrant. Literature data are shown for comparison (INGV 2005). S: spring, W: well

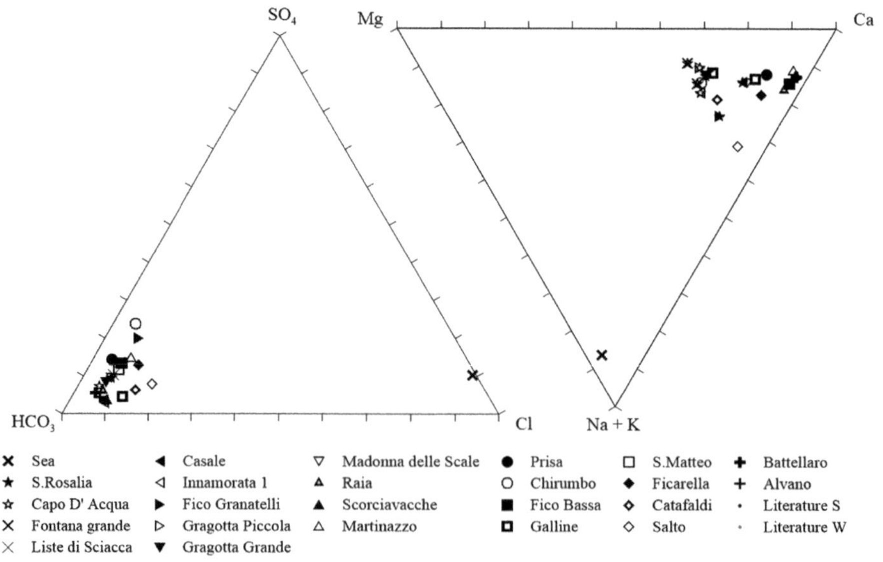

Fig. 10.6 Anion (left) and cation (right) ternary diagrams for Sicani Basin. S: spring, W: well

In addition, the shift towards the magnesium vertex is probably related to the dolomitic or magnesian limestone-rock interaction (Fig. 10.6).

A shift towards the Sulphate vertex would highlight interaction with chalky lithology.

10.4 Isotopic Characteristics of the Catchment Area

The isotopic composition of groundwater varies from −7.7‰ to −5.9‰ for $\delta^{18}O$ and from −42‰ to −32‰ for $\delta^{2}H$ (Table 10.1). Waters from the Sicani hydrogeological basin align on and very close to the LMWL line (Fig. 10.7) determined by Liotta et al. (2013) as sector 2 (see Chap. 1, Table 1.6) confirming a predominant meteoric origin of these waters. The LMWL equation is $\delta^{2}H = 7.67\ \delta^{18}O + 13.9$.

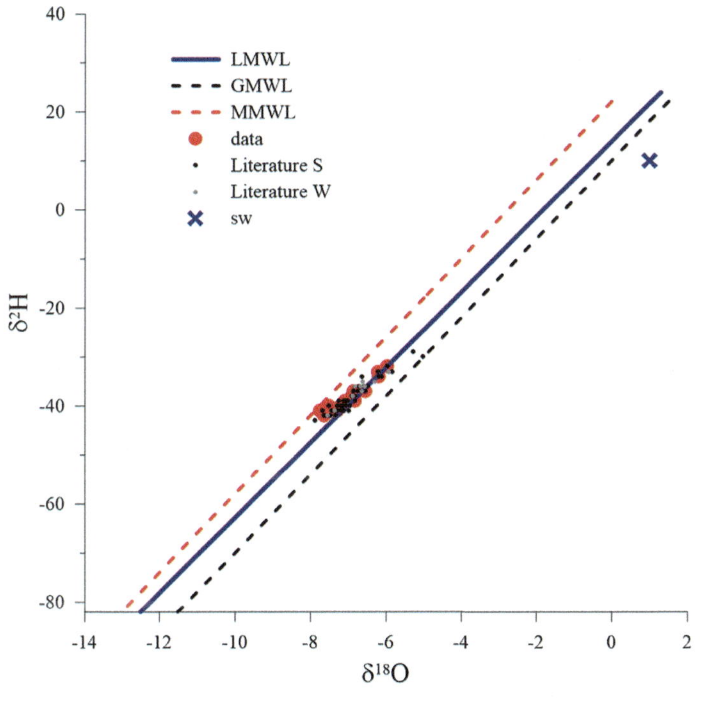

Fig. 10.7 $\delta^{2}H$—$\delta^{18}O$ (in ‰ vs SMOW) diagram of spring waters. The LMWL is from Liotta et al. (2013). The GMWL and MMWL are shown for comparison. Literature data from INGV (2005). S: spring, W: well

According to the isotopic gradient of 0.16δ‰/100 m (Table 1.7 relative to sector 2, Fig. 10.8), the altitude of supply is up to 1.000 m above sea level compatible with the average altitude of the basin.

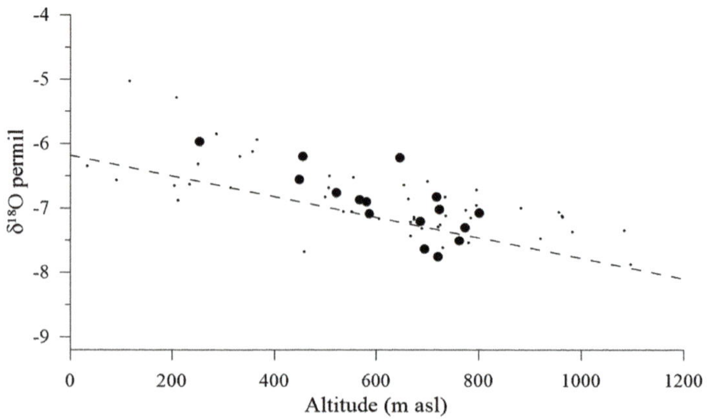

Fig. 10.8 Relationship between $\delta^{18}O$ and altitude

10.5 Trace Elements

In Sicani Mountains Basin, trace element concentrations in water springs are always below the legal limits for drinking water (Fig. 10.9, Tables 1.1 and 10.2).

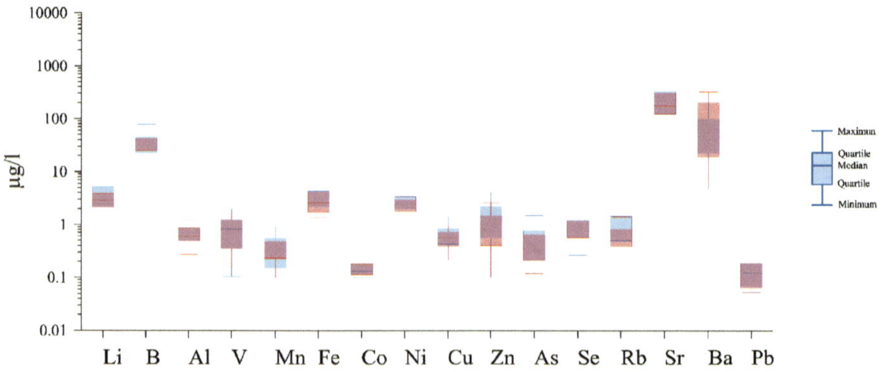

Fig. 10.9 Box-whisker plot of trace elements at Sicani Mts basin. Red box spring data (this work). Light blue box data from INGV (2005)

The range in concentration is wide, and the Sr, Ba and B are, by far, the most concentrated element in all samples (Sr > Ba>>B).

Table 10.2 Trace elements in sampled waters. Concentrations are in μ g/L

	T	EC	pH	Eh	Na	K	Ca	Mg	Cl	NO_3	SO_4	HCO_3	$\delta^2 H$	$\delta^{18} O$	WB
Alvano	14.2	548	7.2	148	0.72	0.04	5.71	0.32	0.7	0.374	0.34	5.45	−37	−6.55	MG
Battellaro	13.9	534	7.31	138	0.66	0.038	5.59	0.28	0.56	0.295	0.44	5.7	−37	−6.86	MG
Salto	18.4	618	7.22	218	1.8	0.113	4.54	0.83	1.74	0.239	0.6	4.5	−32	−5.97	ScM
Catafaldi	18.5	508	7.5	174	0.85	0.056	3.71	1.72	1.3	0.122	0.44	4.35	−33	−6.21	ScM
Figarella	16.3	832	7.12	219	1.43	0.0674	7.5	1.36	1.67	0.777	1.41	6.5	−34	−6.19	ScM
S. Matteo	14.0	355	8.04	144	0.33	0.023	2.41	1.35	0.47	0.017	0.55	3.00	−42	−7.64	SC
Galline	14.6	558	7.38	271	0.71	0.032	4.97	1.28	1.3	0.042	0.37	5.3	−39	−7.02	SM
Fico Bassa	15.6	693	7.08	204	0.98	0.082	7.26	0.47	0.94	0.453	1.26	5.95	−37	−6.76	SM
Chirumbo	13.2	365	7.51	98	0.43	0.031	2.4	1.51	0.36	0.008	1.21	2.86	−40	−7.5	SO
Prisa	12.8	516	7.33	188	0.57	0.055	4.88	1.00	0.46	0.211	1.08	4.8	−39	−7.08	SO
Martinazzo	14.2	546	7.08	194	0.58	0.022	5.5	0.45	0.84	0.2	1.09	4.4	−38	−6.9	SS
Scorciavacche	15.4	357	7.68	216	0.44	0.017	3.79	0.23	0.57	0.317	0.18	3.4	−39	−6.83	SS
Raia	16.0	506	7.75	531	0.83	0.026	5.00	0.39	0.67	0.05	0.46	5.15	−41	−7.29	SS
Madonna delle Scale	14.3	474	7.64	213	0.61	0.032	3.88	1.17	0.62	0.037	0.64	4.5	−39	−7.09	SS
Gragotta Grande	14.1		6.85		0.44	0.026	2.84	1.73	0.51	0.081	0.52	4.2			SW
Gragotta Piccola	14.3		7.35		0.35	0.026	2.79	1.89	0.59	0.081	0.87	4.03			SW
Fico Granatelli	15.9		6.5		1.05	0.051	3.49	1.4	0.68	0.097	1.27	3.62			SW
Innamorata	14.7		6.8		0.31	0.2	2.69	1.65	0.68	0.032	0.17	4.13			SW
Casale	12.6		6.85		0.48	0.02	2.99	0.99	0.59	0.065	0.19	3.79	−40	−7.28	SM
Liste di Sciacca	16.4		6.95		0.48	0.051	2.74	1.81	0.56	0.016	0.63	3.95	−41	−7.68	SW

(continued)

Table 10.2 (continued)

	T	EC	pH	Eh	Na	K	Ca	Mg	Cl	NO_3	SO_4	HCO_3	$\delta^2 H$	$\delta^{18} O$	WB
Fontana grande	12.8		7.1		0.26	0.01	2.2	1.73	0.51	0.048	0.13	3.46	−42	−7.61	SC
Capo D' Acqua	13.4	339	7.28	206	0.36	0.031	2.69	1.61	0.38	0.053	0.41	3.9	−41	−7.75	SC
S. Rosalia	14.5	405	7.65	139	0.38	0.033	3.01	1.68	0.37	0.08	0.43	4.3	−40	−7.2	SC

References

Agnesi V, Macaluso T, Monteleone S, Pipitone G, Sorriso-Valvo M (1989) Types and dynamics of deep gravitational deformations in relation to geological structures. the casestudies of Mt Genuardo and Scopello (western Sicily). Boll Soc Geol It 108:379–389

Ben-Avraham Z, Boccaletti M, Cello G, Grasso M, Lentini F, Torelli L, Tortorici L (1990) Main structural domains originating from the Nogenic-Quaternary collision in the central Mediterranean. Mem Soc Geol It 45:453–462

Boccaletti M, Tortorici L, Ferrini G L (1986) The Calabrian Arc in the frame of the evolution of the Tyrrhenian Basin. In: Boccaletti M, Gelati R, and Lucchi FR (eds) Paleogeography and geodynamics of the Perityrrhenian Area. Giorn Geol 3a, 48(1/2):113–120

Boccaletti M, Ciaranfi N, Cosentino D, Deiana G, Gelati R, Lentini F, Massari F, Moratti G, Pescatore T, Ricci Lucchi F and Tortorici L (1990) Palinspastic restoration and paleogeographic reconstruction of the peri-Tyrrhenian area during the Neogene. Palaeogeo Palaeoclimatol Palaeoecol 77:41–50

Catalano R, Franchino A, Merlini S, Sulli A, Agate M, Basilone L (1998) Materials for the understanding of the deep settlement of west-central Sicily. In: Catalano R, Lo Cicero G (eds) "La Sicilia Occidentale", Field Trip Guide of the 79° Congr. Naz. S. G. I., Mondello (Pa), pp 175–185

Decima A, Wezel FC (1971) Observations on the Sicilian evaporites of central-southern Sicily. Riv Min Sic 130–132:172–187

Flores G (1959) Evidence of slump phenomena (olistostromes) in areas of hydrocarbons exploration in Sicily. Fifth World Petroleun Congr., Sect. 11, Paper 13, New York

Istituto Nazionale di Geofisica e Vulcanologia (INGV), Sezione di Palermo (2005) First characterisation and monitoring of subterranean waters for the drafting of the 'Water Protection Plan for the Region of Sicily'. On behalf of the Sicilian Region - Commissioner for the Waste Emergency and Water Protection. (in italian)

Liotta M, Grassa F, D'Alessandro W, Favara R, Gagliano Candela E, Pisciotta A, Scaletta C (2013) Isotopic composition of precipitation and groundwater in Sicily, Italy. Appl Geoch 34:199–206

Mascle G (1979) Geological study of the Sicani Mountains. Riv It Paleont Strat. Mem 16:1–431

Michetti AM, Brunamonte F, Serva L (1992) Paleoseismological evidence in the epicentral area of the January 1968 earthquakes, Belice, Southwestern Sicily. Ass Eng Geol. Spec Publ 6:127–139

Monaco C, Tortorici L, Catalano S (2000) Tectonic escare in the Sicanian mountains (western Sicily). Mem Soc Geol It 55:17–25

Ogniben L (1954)The Sicilian brecciated clays. Mem Ist Geol Mineral Univ Padova 18:1–92

Ogniben L (1964) Tavejannaz-type sandstones in Sicily. Geol Romana 3:125–170

Ogniben L (1966) Hydrogeological Features of Etna. Riv Min Sic 100–102:1–24

Patacca E, Scandone P, Giunta G, Liguori V (1979) Mesozoic paleotectonic evolution of the Ragusa zone (Southeastern Sicily). Geol Rom 18:331–369

Pedley M, Renda P (1998) A regionally correlatable high to lowstand signal, from late Burdigalian-Early Langhian outliers in Western Sicily. Italy Boll Soc Geol It 117:39–53

Ruggieri G (1959) Preliminary observations on the stratigraphy of the Sciacca region. Riv Min Sic 58–59:189–195

Ruggieri G Sprovieri R (1970) Marls of S. Cipirrello. Illustrative studies of the Geological Map of Italy. Formazioni Geologiche 4:12

Ruggieri G, Torre G (1974) Geology of the areas devastated by the Belice earthquake. Riv Min Sic 139–141:27–48

Ruggieri G, Unti M (1977) The Quaternary of the plateau of S. Margherita Belice (Sicily). Boll Soc Geol It 96:803–912

Schmidt di Friedberg P (1964) Petroleum Lithostratigraphy of Sicily. Riv Min Sic 15(88–90):198–217

Schmidt di Friedberg P, Barbieri F, Giannini C (1960) The geology of the Madonie mountain group (north-central Sicily). Boll Serv Geol It 81:73–140

Sestini G, Flores G (1986) Petroleum potential of the thrust belt and forethroughs of Sicily. A.A.P.G. Mem., 40, pp 567–584

Torelli L, Zitellini N, Argnani A, Brancolini G, De Cillia C, Peis D, Tricart P (1991) Crustal geological section from the Pelagian foreland to the Tyrrhenian back-arc basin (central Mediterranean). Mem Soc Geol It 47:385–399

Vitale FP (1990) Studies on the Belice Valley (Central-Western Sicily). The Plio-Pleistocene foredeep in the framework of the palaeotectonic evolution of the area. PhD thesis, Napoli, 272 pp

Vitale FP (1997) Stacking pattern and tectonics: field evidences from Pliocene growth folds of Sicily (Central Mediterranean). In: Catalano R. (ed) "Time scale and basin dynamics. Sicily, the adjacent Mediterranean and other natural laboratories", field workshop guidebook, Palermo, June 11–13:133–153

Wezel FC (1973) Diachronism of Oligo-Miocene geological events in the Maghrebids. Riv Min Sic 23(142–144):219–232

Wezel FC (1974) Flysch succession and the tectonic evolution of Sicily during the Oligocene and early Miocene. In: Squires CH (ed) "Geology of Italy", Earth Sciences Soc. Libyan Arabian Republic, 105–127, Tripoli

Winnock E (1981) Structure du Bloc Pelagen. In Wezel FC (ed) Sedimetary basin of mediterranean margins. CNR, Italian Project of Oceanography, Urbino, pp 445–464

Trabia and Termini-Imerese Hydrogeological Basin

11

Leonardo La Pica, Sabina Morici, Esterina Gagliano Candela, and Giovannella Pecoraino

11.1 Geological Framework

The Trabia and Termini Imerese mountains fall in the northern sector of the northern Sicilian chain (Fig. 11.1) and are the result of the tectonic superposition of various carbonate, carbonate-silicoclastic and terrigenous geological bodies, of Upper Triassic–Lower Miocene age, with generally southern vergence. These bodies derive from the Miocene deformation of a pelagic basin and escarpment succession (Dominio Imerese Auct.), formed during the Mesozoic distensive phases, crowned by silicoclastic terrigenous deposits (Fysch Numidico Auct.).

Subordinate are the outcrops, reduced to isolated plates, of pelitic and calcareous-marly deposits of the Sicilide Auct. domain.

In the Middle to Upper Pliocene, further deformation of the tectonic build-up is associated with NW–SE transcurrent fault systems, which resulted in the extrusion of geometrically lower tectonic bodies into the structural building. A second NE–SW system, with a transcurrent component, cuts across the previous one and is responsible for some hydrothermal vents (Fig. 11.2).

The structural analysis of the entire tectonic structure of the Trabia Mountains—Termini Imerese basin has made it possible to recognise a stack of superimposed tectonic units, arranged in this way, from bottom to top:

L. La Pica · S. Morici · E. Gagliano Candela (Deceased) · G. Pecoraino (✉)
Istituto Nazionale di Geofisica e Vulcanologia, Sezione di Palermo, Via Ugo La Malfa 153, 90146 Palermo, Italy
e-mail: giovannella.pecoraino@ingv.it

L. La Pica
e-mail: leonardo.lapica@ingv.it

S. Morici
e-mail: sabina.morici@ingv.it

E. Gagliano Candela et al. (eds.), *Drinking Water in Sicily*,
https://doi.org/10.1007/978-3-031-60473-7_11

Fig. 11.1 The Trabia-Termini Imerese hydrogeological basin. Imagine from Google Earth Pro v 7.3.6.9345 (December 29, 2022). Sicily, Italy

Imerese Units

Pizzo Chiarastella Unit

It is the lowest unit in the structural build-up and rests directly on the Trapanese carbonate platform units outcropping in the Bolognetta area. It is dissected by NW–SE transverse faults that have determined the extrusion of the relief.

Pizzo Di Cane-Monte S. Calogero Unit

This unit is an anticlinal ridge, repeatedly dissected by distensional and/or transcurrent tectonics. As a whole, the unit consists of a ramp anticline verging from SW to SE. Near the front of the ramp, there are calcareous or calcareous-silicicoclastic plates, which are the product of collapses and/or the sliding of blocks or megablocks (Cozzo del Leone, Cozzo Ciauli, near Ventimiglia, Cozzo S. Giovanni near Caccamo and most probably also Mt Castellaccio near Sciara). Low flow springs are associated with some of these blocks. In correspondence with Mt S. Calogero, the anticlinal structure is ripped by direct faults plunging towards the sea, with rejections often exceeding 1000 m and responsible for the springs.

Mount S. Onofrio—Mount Rotondo Unit

This unit is constituted by a ramp anticline, even more dissected by distensive and transcurrent tectonics than the previous one. The southern flank of the structure

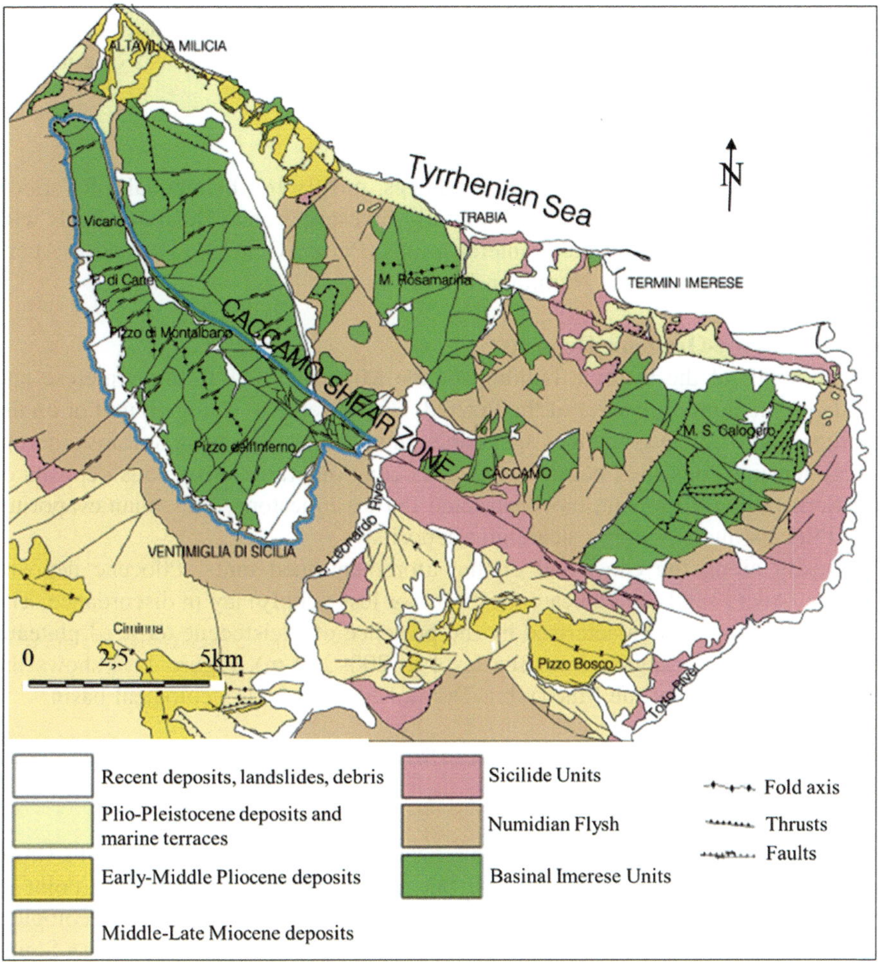

Fig. 11.2 Geological scheme of the Trabia-Termini Imerese Mountains hydrogeological basin (modified after Guarnieri 2004)

tends to become clearly vertical towards the N, and in places to overturn (Cozzo Cannemasche).

Mount Rosamarina—Mount Pileri Unit

This unit is characterised by the presence of inverse faults directed towards the S-SE. In the Mount Pideri—Cozzo Pileri sector, several superimposed tectonic faults are recognisable, sometimes involving the Numidian Flysch, wedged along the overlapping fronts.

Capo Grosso—Torre Colonna Unit
This is the highest of the Imerese units that make up the vast majority of the tectonic structure of the Trabia-Termini Imerese mountains and consists of a number of superimposed sigmoidal flakes with a SE to SW gradient.

Sicilide Unit
They consist of more or less extensive plates, mostly brachisinclinal, of Varicolored Clays Auct. and of the overlying marly limestones and calcareous marls with biocalcarenitic or biocalciruditic intercalations (Polizzi Generosa Formation Auct.) resting tectonically on the Numidian Flysch.

Late- and Post-Orogenic Deposits
To the south of the tectonic building of the Mts of Trabia-Termini Imerese and partly on the numidic covers of the Pizzo di Cane-Monte S. Calogero unit or on the Sicilids, some brachyssynclines are visible, in which late-orogenic terrains outcrop. These terrains consist of pelitic-arenaceous-conglomeratic terrigenous deposits of the Serravallian-Tortonian, bioconstructed Porites limestone, Messinian evaporites and Pliocene marly-sandy-calcarenitic deposits.

Mostly along the coast and on the aforementioned units, Pliocene deposits (*'Trubi'* Auct., sands and calcarenites more or less pebbly) lay in discordance. The coastal strip is also characterised by the presence of Pleistocene terraced plateaus in several orders, ranging from sea level to 200–250 m). Figure 11.2 shows the geological scheme of the Mts Trabia-Termini Imerese hydrogeological basin.

11.2 Hydrogeology

The Trabia-Termini Imerese Mts are of considerable interest not only from a stratigraphic and structural point of view, but also from a hydrogeological point of view: the high values of annual rainfall and the particular existing hydrogeological structures make this sector strategic for the water supply in central-northern Sicily.

The structural complexity of the Trabia-Termini Imerese Mts has considerable implications on the water circulation within the hydrogeological structures of this area. Some authors (Bartolomei et al. 1983; Contino et al. 2004a, b, c) believe that the thermal springs, some of which are submarine, in the area of the Rocca del Castello di Termini, given the small extension of the outcrop, must be fed by a deeper circuit that is most likely related to Mount S. Calogero (Fig. 11.3).

The main hydrogeological complexes are:

Calcareous-Marly and Clay-Arenaceous-Calcareous Complex
It is essentially represented by variegated mudstones and marls with intercalated calcilutitic, calcisiltitic, calcarenitic levels and dolomitic breccias and, in the Cozzo Pideri sector (Contino 2002), by claystones, quartzose-micaceous marls and sandstones, calcareous breccias and organogenic calcareous breccias (Formation of Mufara Auct., Carnico). This complex, as a whole, has a very low degree of

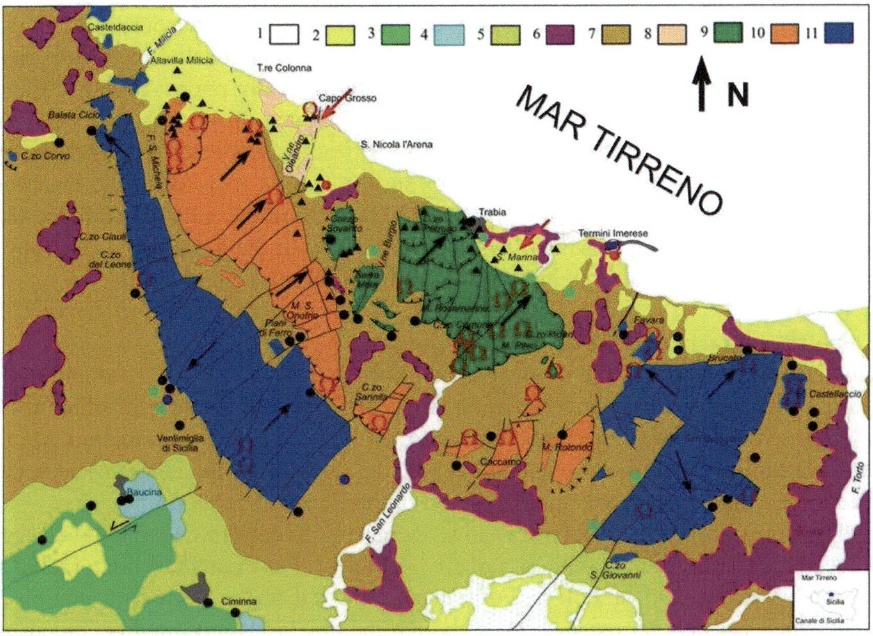

Fig. 11.3 Hydrogeological scheme of the Trabia-Termini Imerese Mountains (modified after Contino et al. 2015). (1) Alluvial deposits (Holocene) (2) Carbonatic and clastic deposits (Pliocene—Pleistocene) (3) Evaporites (Messinian) (4) Recifal limestone (Messinian) (5) Clastic-terrigenous deposits (Serravallian-Tortonian) (6) Sicilidy Units (7) Numidian Flysch (8) Capo Grosso-Torre Colonna Units (Trias-Eocene) (9) Mt Rosamarina-Mt Pileri Unit (Trias-Eocene) (10) Mt S. Onofrio-Mt Rotondo Unit (Trias-Eocene) (11) Mt Cane-Mt S. Calogero Unit (Trias-Eocene). Green circles: drinkable water wells; Black circles: springs

permeability, which is further reduced when it is intensely tectonised. Permeability is between 10^{-7} and 10^{-9} m/s. The degree of vulnerability is low, practically negligible.

Calcareous-Dolomitic Complex

Various lithological terms, permeable mainly by fracturing, with a similar degree of permeability have been grouped in this complex. These are calcilutites and calcarenites of the Scillato Formation, dolomitic breccias, dolorudites and doloarenites of the Fanusi Formation, and large resedimented carbonate bodies (essentially calcarenites and calcirudites) of the Crisanti Formation. Due to its hydrogeological characteristics, this complex represents one of the main aquifers of this area, as attested by the presence of some important springs located to the north of Mt S. Calogero.

The fracturing develops in several systems, generally the main one presents a trend almost perpendicular to the secondary one and even if it is not accompanied by particularly developed karst phenomena, it makes these terms of excellent aquifers with high permeability from 10^{-1} to 10^{-3} m/s. The considerable exposure at the outcrop allows for a large supply that generates high flow sources.

The prevailing permeability is high due to fracturing and locally due to porosity, when it is intensely cataclysmic. Karst forms are generally poorly developed and are often linked to tectonic discontinuities. The most conspicuous underground water tables of the di Trabia-Termini Imerese Mts area are located in this complex and feed large springs (e.g. the Brucato spring group). This complex emerges mainly in the two ridges of Mt Cane- Mt S. Onofrio, in the mountains above the built-up area of Trabia, in Mt S. Calogero and partly in the Rocca del Castello of Termini Imerese, the latter being the seat of hydrothermal circulation.

Calcareous Marly and Silico-Marly Complex

This complex comprises the sequences of siliceous argillites, radiolarites and radiolarian marls of the Crisanti Formation Auct., the calcilutites and calcisilites with intercalated banks of graded biocalcarenites of the Caltavuturo Formation Auct., the calcilutites, calcisilites and marls with flint strips ("Scaglia" Auct.) and the calcilutites, calcarenites and marls of the Polizzi Formation Auct.

The permeability by cracking is low, increasing locally in calcarenite successions. These are terrain in which porosity is almost non-existent, while cracking, especially in the case of Polizzi Formation, is linked to the degree of tectonisation, which is often very strong, even though the diaclasms are often occluded by silty-lutitic material. Permeability is however on average contained within 10^{-5} to 10^{-7} m/s. Vulnerability is medium to low.

Meso-Cenozoic Carbonate Megabreccia Complex

These are mainly calcareous megabreccias and olistolites in lenticular or stratiform bodies with a thickness varying greatly from a few centimetres to several tens of metres and interspersed with various levels in the clayey successions of the Numidian Flysch. Porosity is almost nil, while cracking is linked to their degree of tectonisation; permeability varies from 10^{-1} to 10^{-2} m/s. The modest outcrops generate small springs of ephemeral flow. Vulnerability is on average medium, but high when extremely localised.

Numidian Flysch Arenaceous Clay Complex

The pelitic-arenaceous sequences of the Numidian Flysch are represented in the complex. These are sandstones with intercalations of clays and conglomeratic sandstones. Permeability is essentially by cracking in correspondence with the arenaceous levels. Permeability is variable depending on the state of fracturing of the sandstone beds. The prevalence of arenaceous terms over pelitic terms leads to a subterranean water circulation that, although discontinuous, is on the whole of some interest given the frequency of the fracturing planes. This is evident from a whole series of small springs that border, at different heights, the arenaceous beds above the poorly permeable terms or within the same succession where there are more extensive clay intercalations. Permeability is between 10^{-7} and 10^{-9} m/s. Vulnerability is low and practically negligible.

Predominantly Arenaceous Complex

The conglomeratic sandstone sequences of the Numidian Flysch are grouped in this complex. This complex is of some hydrogeological interest in local terms, since the number of springs is always linked to the extension of the outcrops and the frequency and thickness of the sandstone beds, which are generally very fractured. In correspondence with the latter, the permeability is generally discrete but can be high in some cases. Permeability is modest due to porosity, locally higher due to the presence of diffuse cracking; average values are close to 10^{-5} m/s.

Predominantly Clayey-Marly Complex

The lithological and hydrogeological affinities of this complex bring together the numerous formations belonging to the Sicilide and Oligo—Miocene Units. In fact, the following are represented in this complex: the Serravallian—Tortonian clays (Pre-Terravecchia Auct.), the marls and sandy clays of the Tortonian (Terravecchia Auct. Formation), the thinly stratified clays, siltstones and sandstones of the Numidian Flysch (pelitic facies), the Varicolored Clays Auct., The Varicoloured Clays Auct., frequently tectonised, sometimes chaotic, and incorporating a more or less abundant stone fraction, the alternating arenaceous and clayey-marly levels of the Tusa volcanoclastites (Tusa Formation or 'Tufiti di Tusa' Auct.). Permeability is low to very low, with values between 10^{-8} and 10^{-9} m/s. Vulnerability is practically zero, except in localised situations, due to the presence of small springs or wells.

Conglomeratic Complex

Polygenic conglomerates, mostly pebbly, more or less cemented, with predominantly carbonate and/or arenaceous clasts and a sandy matrix are found in the southern sector of the Trabia-Termini Imerese Mts. These conglomerates are usually poorly stratified with sandy levels and lenses, sometimes sandy-clayey, especially at the base. Permeability is generally low, ranging from $10^{-7}$to 10^{-5} m/s. Vulnerability is considered medium to low, with local conditions of greater sensitivity to point source pollution.

Calcareous Chalky-Marly Complex

This complex includes all the terms pertaining to the evaporitic series and the marly and marly-calcareous terms of the *'Trubi'*. In this complex are grouped: marls and calcareous marls (*'Trubi'*), gypsumarenites alternating with banks of macrocrystalline gypsum with clay intercalations, calcarenites and organogenic limestones biolithic at Porites (Baucina Auct. Formation), lacustrine clays with muddy conglomerate intercalations ('Fanglomerates'). Relative permeabilities vary depending on the physical state, such as the state of karst dissolution of chalks. In relation to this, medium to low permeability can be assigned by porosity and subordinately by cracking.

The waters that this complex contains are mostly highly mineralised, and generally more or less of poor quality and therefore not usable for drinking water purposes.

The outcrops of the 'Porites limestones' of the Baucina Formation, the Messinian terms and the overlying Pliocene *'Trubi'* are most widespread in the so-called Ciminna Basin.

Permeability is low for porosity, while it acquires a certain relevance where limestone and/or gypsum cracking is more accentuated; it is however between 10^{-5} and 10^{-7} m/s. The vulnerability of the evaporitic complex is considered to be medium to low; it is more relevant where chalk and/or limestone are present in the outcrop, and lesser or almost negligible where the *'Trubi'* mask the underlying evaporitic terms.

Alluvial Complex and Littoral Deposits

The alluvial complex, outcropping in the valley floor and along the river mouths, consists of lenticular bodies of polygenic materials, from coarse to fine. The repeated presence of silty-clay lenticular bodies, more frequent in the river stretches with low inclination and towards the coast, determines the separation of levels with different permeability and thickness. The presence of levels with different degrees of permeability therefore allows us to define this complex as a multi-stratum aquifer, since the water table contained in these materials is irregularly divided into a series of superimposed bodies of water whose piezometric height is quite variable.

Permeability can vary from 10^{-3} to 10^{-7} m/s. The vulnerability of alluvial aquifers is high due to their high average permeability and to the anthropogenic exploitation conditions to which they are subjected, especially on the valley floors and along the coast.

Aquifer Detritus and Alluvial Coverings

Detrital materials and sometimes eluvial products and colluvial deposits are present along the flanks of the main carbonate reliefs and at the foot of the thick quartzarenitic banks of the Numidian Flysch. The detrital materials are essentially made up of heterometric clasts, even of large dimensions. Occasionally thin lenses of material derived from soil runoff are present. Permeability is generally high, from 10^{-3} to 10^{-5} m/s and drainage is very rapid. The hydrogeological role is to drain and distribute in depth the infiltration waters and those drained by the reliefs to which they are adjacent.

Within the Trabia-Termini Imerese Mts hydrogeological basin, the following five hydrogeological units may be distinguished (Fig. 11.4):

- Pizzo Chiarastella hydrogeological unit
- Pizzo di Cane-Mt S. Calogero hydrogeological unit
- Mount S. Onofrio-Mount Rotondo hydrogeological unit
- Mount Rosamarina-Mount Pileri hydrogeological unit
- Capo Grosso-Torre Colonna hydrogeological unit

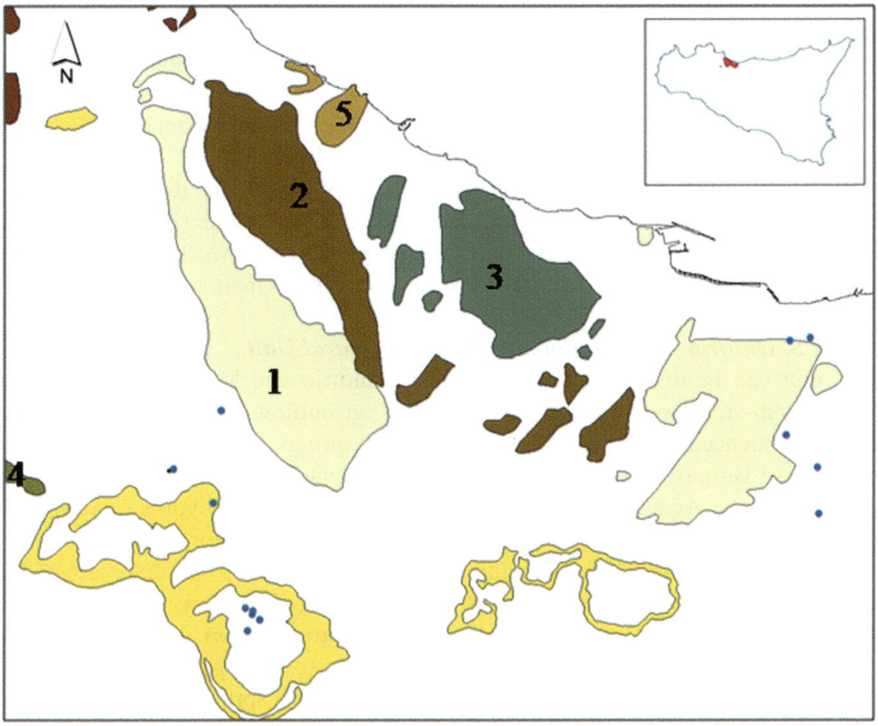

Fig. 11.4 Map of Mt of Trabia-Termini Imerese hydrological basin. Numbers identify water bodies: (1) Pizzo Cane-Mt S. Calogero; (2) Capo Grosso-Mt Colonna; (3) Mt Rosamarina-Mt Pileri; (4) Pizzo Chiarastella; (5) Capo Grosso-T. Colonna. Location of spring is shown (green plots represent PRGA springs, INGV 2005, blue dots indicate springs sampled for this study)

Pizzo Chiarastella Hydrogeological Unit

This hydrostructure has a main outflow towards the NW towards the Bagni di Cefalà thermal spring. In addition, it is exploited by thermal wells with little mineralised water used for drinking water purposes (Municipality of Villafrati). The free water table is hosted in the network of fractures of the calcareous complex (Scillato Formation Auct.).

Pizzo Di Cane—Monte S. Calogero Hydrogeological Unit

This hydrostructure can be divided into the substructures of Pizzo di Cane to the west and Monte S. Calogero to the east. The former has a main outflow to the northwest, as evidenced by the springs at Granatelli and Masseria Navurra; the latter, instead, is used for irrigation and drinking water via the Casteldaccia municipal aqueduct. The substructure of Monte S. Calogero is ripped by faults directly immersing towards the N, with displacement exceeding a thousand metres, responsible for the manifestations of springs, due to the superimposed permeability threshold, and the outflow is mainly directed towards the N-NE (Brucato springs today captured by wells, Fontana Fredda).

Other springs linked to this sub-structure, due to their superimposed permeability threshold, are those of Favara-Scamaccio, south of the built-up area of Termini Imerese, which flow from the Cretaceous limestone (Crisanti Formation Auct.) and the chloride-salt-iodine thermal springs of Termini Imerese.

In this hydrostructure, the main aquifer is represented by the two overlying calcareous (Scillato Formation Auct.) and calcareous-dolomitic (Fanusi Formation Auct.) complexes, which host a water table in a fracture network, free or partially confined by the overlying radiolarites of the Crisanti Formation Auct.

Monte S. Onofrio—Monte Rotondo Hydrostructural Unit

This unit can be divided into the Monte S. Onofrio and Monte Rotondo-Cozzo Sannita sub-structures. The first one presents an outflow mainly towards the E–NE, as evidenced by the modest or ephemeral springs located along the detrital belts (Case Burgio, Roveta, Speciale), but also towards the SE, as attested by the presence of the Acquagrande spring group. From E-NE the hydrostructure feeds the overlying unit (tectonic contact) of Monte Rosamarina-Monte Pileri.

On the western side of the structure, along the overthrust front, there are modest springs (e.g. Piani di Ferro), some of which are linked to landslide megablocks, such as those of Case Randino (Randino and S. Pietro springs). Towards the north, the sub-structure tends to verticalize and in places to overturn, giving rise to the Acqua Papa spring in the Iencherìa district. Acqua Papa spring in 1929 had a flow rate of 15 l/s, which dropped to 2 l/s at the end of the 1970s and has now become very low after the construction of the Bucaro I, II and III wells (total flow rate of about 200 l/s). The Bucaro wells currently supply the municipalities of Ventimiglia and Ciminna.

Similarly to the previous unit, the main aquifer is represented by the two superimposed calcareous (Scillato Formation Auct.) and calcareous-dolomitic (Fanusi Formation Auct.) complexes, which host a fracture network aquifer, free or partially confined by the overlying radiolarites of the Crisanti Formation Auct.

Monte Rosamarina-Monte Pileri Hydrogeological Unit

This unit can be divided into the sub-structures Cozzo Sovarito, Serra Mola, Cozzo-Malpertuso-Pizzo Camercia, Cozzo Comune-Monte Rosamarina-Monte Pileri. The modest springs linked to this hydro-structure have undergone a considerable decrease in outflow over the last 50 years, so much so that they have practically become scarce or even extinct.

The Cozzo Malpertuso-Pizzo Camercia sub-structure was characterised by the presence of the Acqua dell'Oro (Acqualoro) spring group, which was very productive, with flow rates of over 100 l/s, and used for drinking water. In order to compensate for the considerable decrease in flow rate suffered by this spring over the last 50 years (to which the construction of the Palermo-Catania motorway and a number of private wells does not seem unrelated), in the 1970s the Trabia local authority constructed the Fornace, Camercia, Campo Sportivo I and II and Trabia wells (the latter unused).

The main aquifer is once again represented by the two superimposed calcareous (Scillato Formation Auct.) and calcareous-dolomitic (Fanusi Formation Auct.) complexes, which host a water table in a fracture network, free or partially confined by the overlying radiolarites of the Crisanti Formation Auct.

Capo Grosso-Torre Colonna Hydrogeological Unit

The flake structure of this unit is attested by the stratigraphy of the E.S.A. (Ente Siciliano Acque) wells 'S. Michele' and 'Piano Aci II', which intercepted tectonic doubling in the succession of the terrain. The Acqua Calda spring, used for irrigation purposes, is connected to the NE–SW system of transcurrent faults, mainly left-handed, that carve out the Trabiese tectonic structure.

The hydrostructure is exploited for wells and has several springs along the coast or underwater, since the outflow is oriented towards the sea. It is likely to receive a certain amount of water from the Monte S. Onofrio—Monte Rotondo hydrostructure. The main aquifer is represented by the two superimposed calcareous (Scillato Formation Auct.) and calcareous-dolomitic complexes (Fanusi Formation Auct.), which host a water table in a fracture network, free or partially confined by the overlying radiolarites of the Crisanti Formation Auct.

11.3 Hydrogeochemical Characterisation

Two samples belonging to the Pizzo di Cane-Monte S. Calogero water bodies were sampled (Table 11.1). The examined waters samples are characterised by similar T and pH (between 19 and 22 °C and pH ~ 7.7), but differ in EC, although it is low in both samples (528 and 121 μS/cm).

Most of the spring waters, both sampled for this study and literature data (INGV 2005), fall in the bicarbonate-alkaline-earthy field of the Langelier-Ludwig diagram (Fig. 11.5). Also considering the literature data (INGV 2005), one sample falls into the chloride-sulfate-alkaline and the chloride-sulfate-alkaline-alkaline-earthy range, assuming, therefore, that there may be a mixing of waters between the two extreme terms or evolution from bicarbonate-alkaline-earthy waters to chloride-sulfate-alkaline waters.

Table 11.1 Chemical and isotopic composition of the sampled points

	T	EC	pH	Eh	Na	K	Ca	Mg
Brocato	19.6	528	7.75	375	0.93	0.053	2.55	2.5
Favara	21.8	121	7.72	463	0.71	0.068	3.58	1.16
	Cl	NO_3	SO_4	HCO_3	δ^2H	$\delta^{18}O$	WB	
Brocato	0.77	0.027	0.69	4.42	−44	−8.14	PC-MSC	
Favara	0.50	0.048	0.98	3.85	−36	−6.53	PC-MSC	

Analytical data are expressed as meq/L, EC in μS/cm and δ^2H and $\delta^{18}O$ in ‰ versus SMOW. WB is for water body. PC-MSC is for Pizzo di Cane-Mt S. Calogero water bodies

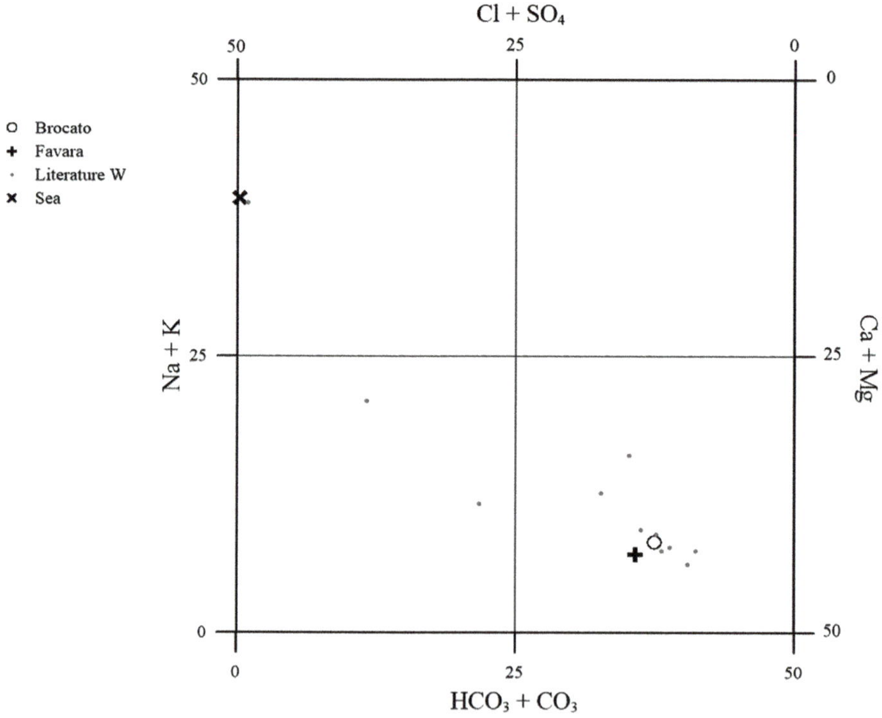

Fig. 11.5 Langelier–Ludwig classification diagram. The springs fall in the bicarbonate-alkaline-earthy quadrant. Literature data are shown for comparison (INGV 2005). W: well

This characteristic is also evident in Fig. 11.6, where it is shown that the waters have a well-defined calcium-magnesium ratio, compatible with the type of rocks that make up the main lithofacies of the Mt of Trabia and Termini Imerese.

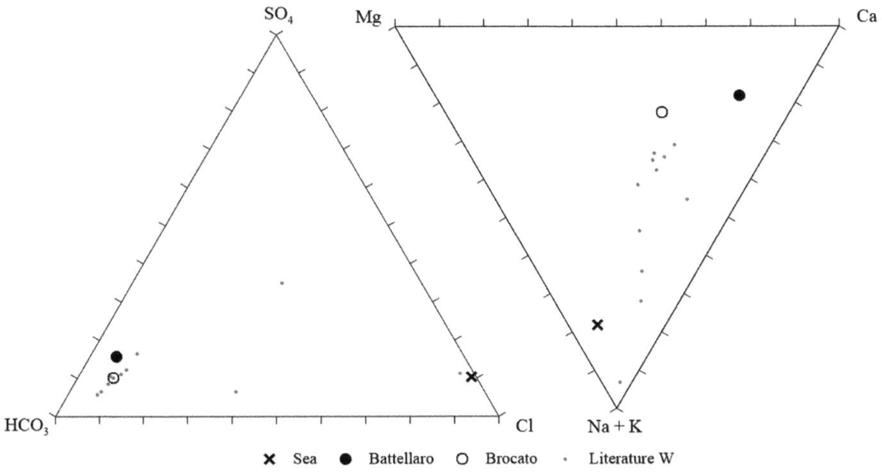

Fig. 11.6 Anion (left) and cation (right) ternary diagrams for Trabia-Termini Imerese Basin. W: well

There is also a trend towards an increasing sodium component, in general, as the waters are sampled towards the coast related to seawater contamination. The trend towards the sulfate vertex (Fig. 11.6 left) suggests circulation in evaporitic lithologies.

11.4 Isotopic Characteristics of the Catchment Area

Sampled waters show $\delta^{18}O$ and $\delta^{2}H$ isotopic composition ranging from -8.1 to -6.5, and from -44 to -36 δ‰ versus SMOW, respectively, aligning very close to the MMWL (Fig. 11.7). Points plot on the local meteoric water line determined by Liotta et al. (2013) as sector 1 (see Chap. 1, Table 1.7; $\delta^{2}H = 6{,}31\delta^{18}O + 5$) confirming a meteoric origin for these waters. The strong difference in composition is probably due to the different altitude of emission and alimentation.

The isotopic gradient of 0.15δ‰/100 m (Table 1.8 relative to sector 1) and the isotopic composition more negative respect to meteoric water suggest a supply altitude higher than 1000 m, compatible with the average altitude of the basin (Fig. 11.8).

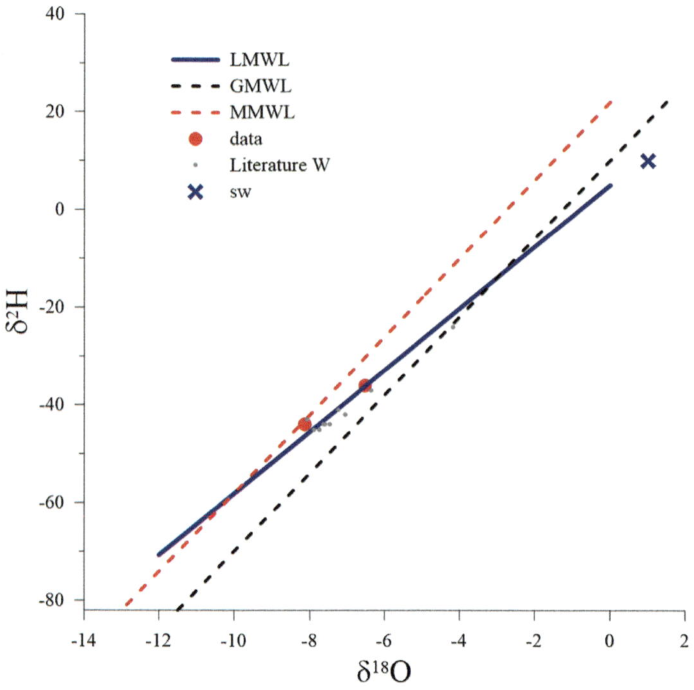

Fig. 11.7 δ^2H–$\delta^{18}O$ (in ‰ versus SMOW) diagram of spring waters. The LMWL is from Liotta et al. (2013). The GMWL and MMWL are shown for comparison. Literature data from INGV (2005). W: well

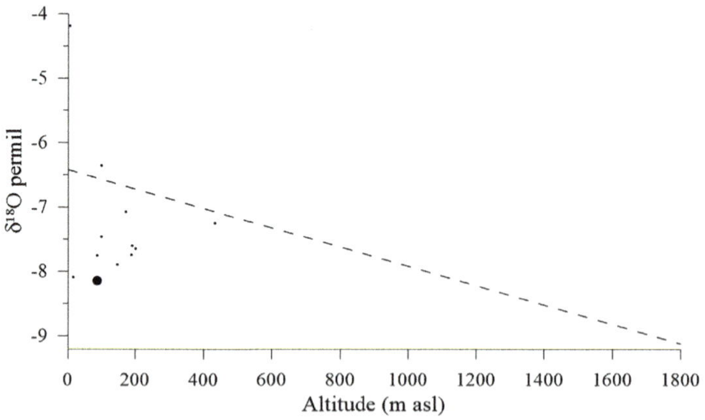

Fig. 11.8 Relationship between $\delta^{18}O$ and altitude

Bibliography and Recommended Further Reading

Abate B, Renda P, Tramutoli M (1988) Illustrative notes of the geologic map of the Termini Imerese and Western Madonie Mountains (north-central Sicily). Mem Soc Geol It 41:475–505

Alaimo R, Carapezza M, Dongarrà G, Hauser S (1978) Geochemistry of Sicilian hot springs. Rend Soc Chim Ital 34(2):577–590

Catalano R, Montanari L (1979) Geology of the Trabia-Termini Imerese Mountains and the eastern Sicanian Mountains (Bagheria and Termini Imerese Sheets. North-central Sicily). Rend Acc Sc Fis e Mat Napoli 46:1–29

Contino A, Di Maio D, Lo Cicero G, Scozzola N (2002) Geomorphological aspects of the Gulf of Termini Imerese. 81st Summer Meeting of the Italian Geological Society, Torino 10–12 september 2002 pp 106–107 (Abstract)

Contino A, Cusimano G, Frias Forcada A (2004a) Hydrostructural layout of the Trabia and Termini Imerese Mountains (Northern Sicily) and evaluation of stored water resources. "Exploratory Activity and Research in Sicilian Karst Areas," Proceedings of the 4th Conference of Speleology of Sicily, 1°–5 May 2002, Custonaci (TP). Suppl. Speleologia Iblea, 10:99–106, a.2002. Ragusa. Pubblicazione n. 2595 del G. N. D. C.—C. N. R., U. O. 4.17

Contino A, Cusimano G, Frias Forcada A, Gatto L, Hauser S (2004b) Hydrogeological resource potential and quality in the Trabia and Termini Imerese mountain hydrostructures (Southern Sicily)—32nd International Geological Congress, Florence, Italy, August 20–28, 2004. Accepted as "Poster Presentation" in session T 38.01 "Groundwater management in vulnerable areas"

Contino A, Cusimano G, Gatto L, Hauser S, Pisciotta A (2004c) The coastal aquifers of the Trabia-Termini Imerese Mountains: modes of water circulation, role of karstification and vulneration by marine ingression phenomena. 2nd International Seminar of Studies on Karstification in the Hyblaean and South Mediterranean Area, 28–30 may 2004, Castello di Donnafugata (RG), volume summaries, 25

Contino A, Monteleone S, Sabatino M (2015) Water resource assessment in karst and fractured aquifers of termini Imerese-Trabia Mts. (Northern Central Sicily, Italy). In: Lollino G et al (eds) Engineering geology for society and territory, vol 3, pp 573–577. Springer, New York. https://doi.org/10.1007/978-3-319-09054-2_115

Contino A (2002) Geology and Geomorphology of the Lower Northern Imera River Valley. In: Belvedere O, Bertini A, Boschian G, Burgio A, Contino A, Cucco RM, Lauro D, Himera D (eds) Archaeological Prospecting in the Imera Valley. Department of Historical-Archaeological, Socio-anthropological and Geographical Cultural Heritage, Archaeology Section, University of Palermo pp 25–48, Libreria Editrice L'Erma di Bretschneider, Roma, p 456. Publication as part of the project "Archaeological map of Italy, applied cartography in urban and territorial areas." Attached is a geological and geomorphological map of the lower valley of the northern Imera River (Map of geological and geomorphological elements), scale 1: 25000

Cusimano G, Liguori V (1988) Hydrodynamics and potential of groundwater resources of the Trabia-Termini Imerese Mountains hydrogeological system (Sicily). Underground Waters 5:49–63

Guarnieri P (2004) Structural evidence for deformation by block rotation in the context of traspressive tectonics, northwestern Sicily (Italy). J Struct Geol 26(2):207–219

Istituto Nazionale di Geofisica e Vulcanologia (INGV), Sezione di Palermo (2005) First characterization and monitoring of groundwater aimed at the drafting of the "Water Protection Plan of the Sicily Region". On behalf of the Sicilian Region—Deputy Commissioner for the waste emergency and water protection

Kiratzi AA (1994) Active seismic deformation in the Italian Peninsula and Sicily. Ann Geof 37(1):2–4

Liotta M, Grassa F, D'Alessandro W, Favara R, Gagliano Candela E, Pisciotta A, Scaletta C (2013) Isotopic composition of precipitation and groundwater in Sicily, Italy. Appl Geoch 34:199–206

Lo Cicero G, Di Stefano E, Catalano R, Sprovieri R, Agate M, Contino A, Greco G, Mauro G (1997) The Ciminna Messinian basin cyclical sedimentation and eustatic control in a transpressive tectonic setting. In: Catalano R (eds) Guidebook, field in western Sicily, 8th workshop of the ILP task force "origin of sedimentary Basins", Palermo (Sicily) June 7–13, 1997, pp 71–86
Marchetti MP (1960) Summary introduction to geology of Sicily. Petrol Explor Soc Lybia, Excursion in Sicily 47–60
Miuccio A, Frixa A, Bertamoni M (2000) The trapanese structural domain in the termini imerese mountain area (Sicily). Mem Soc Geol It 55:227–234
Montanari L (1968) Material for the geology of the upper Lercarese (Sicily). Boll Soc Geol It 87:133–142
Montone P, Amato A, Frepoli A, Mariucci MT, Cesaro M (1997) Crustal stress regime in Italy. Ann Geof 40(3):741–757
Nigro F, Renda P (2000) A model of tectono-sedimentary evolution of the Sicilian Neogene foreland. Boll Soc Geol It 119:667–686
Pepe F, Bertotti G, Cella F, Marsella E (2000) Rifted margin formation in the south Tyrrhenian Sea: a high-resolution seismic profile across the north Sicily passive continental margin. Tectonics 19(2):241–257
Rangin C (1975) Contribution to the paleogeographic and structural study of the Termini Imerese Mountains (Sicily). Bull Soc Geol France 17:38–44
Ruggieri G (1966) Notes on the Miocene of western Sicily. Riv Min Sic 97–99:18–24
Schimdt di Friedberg P (1964) Petroleum Lithostratigraphy of Sicily. Riv Min Sic 15:198–217; 16:50–71
Schmidt di Friedberg P, Barbieri F, Giannini G (1960) The geology of the Madonie mountain range (north-central Sicily). Boll Serv Geol It 81:73–140
Schmidt Di Friedberg P (1962) Introduction to the petroleum geology of Sicily. Rev Inst Fr Petr Ann Comb Liq 17:635–668
Schutte KG (1978) Crustal structure of southern Italy. In: Closs H, Roeder D, Schmidt K (eds) Alps. Apennines, Hellenides, Stuttgart, pp 315–321

Trapani Mountains Hydrogeological Basin

12

Giovannella Pecoraino, Leonardo La Pica, Esterina Gagliano Candela, and Sabina Morici

12.1 Geological Framework

The Trapani Mountains are located at the western end of Sicily and are bounded to the north by the Tyrrhenian Sea, to the east by the Palermo Mountains, to the south by the so-called Castelvetrano Basin and to the west by the Strait of Sicily (Fig. 12.1). The most important reliefs are Mt Sparagio and Mt Inici, characterised by altitudes around 1.000 m (1.100 and 980 m, respectively). The Trapani Mts are formed by the Miocene tectonic superposition of various carbonate, carbonate-silicicoclastic and terrigenous units, dating from the Upper Triassic to the Upper Miocene, directed towards the S and SE (Giunta and Liguori 1970, 1972; Catalano and D'Argenio 1982; Abate et al. 1991, 1993; Fig. 12.2).

These units derive from the deformation of carbonate and carbonate-pelagic platforms identified during the Mesozoic distensive phases. The geometrically lower units derive, according to the most recent structural models, from the deformation of the soils of the Trapanese-Saccense Dominion, which are superimposed on the units deriving from the deformation of the westernmost sector of the Panormide Dominion. The geometrically higher units derive from the deformation of Cretaceous-Neogene terrains detached from their Mesozoic substrate and referred to the Pre-Panormid units. The outcropping terrains are mostly carbonatic.

G. Pecoraino (✉) · L. La Pica · E. Gagliano Candela (Deceased) · S. Morici
Istituto Nazionale di Geofisica e Vulcanologia, Sezione di Palermo, Via Ugo la Malfa 153, 90146 Palermo, Italy
e-mail: giovannella.pecoraino@ingv.it

L. La Pica
e-mail: leonardo.lapica@ingv.it

S. Morici
e-mail: sabina.morici@ingv.it

E. Gagliano Candela et al. (eds.), *Drinking Water in Sicily*,
https://doi.org/10.1007/978-3-031-60473-7_12

Fig. 12.1 The Trapani Mountains hydrogeological basin. Imagine from Imagine from Google Earth Pro v 7.3.6.9345 (December 29, 2022). Sicily, Italy

The following successions have been recognised on the basis of their lithostratigraphic characteristics and geometric position:

1. Meso-Cenozoic carbonate and silicoclastic successions with carbonate and carbonate-pelagic platform characteristics referable to the Trapani Domain;
2. Meso-Cenozoic carbonate successions with carbonate platform, margin and slope terrigenous characteristics, referable to the Panormide Domain (Catalano and D'Argenio 1982; Abate et al. 1993);
3. Carbonate and silicoclastic successions with basin and platform margin characteristics referable to the Pre-panormid Dominion from the Cretaceous to the Miocene (Catalano et al. 1996);
4. Late- and post-orogenic deposits discordant with the successions above described, consisting of terrigenous, evaporitic, silicoclastic and carbonate deposits of the Upper Tortonian—Pleistocene interval.

The structural set-up of the area is the result of the interaction of the tectonic phases that occurred from the Miocene to the Pliocene. Three groups of superimposed tectonic units can be distinguished, with ramp-flat geometries and generally directed towards the southern areas. The lowest units are those that derive from the deformation of the Trapanese Dominion, which are in turn tectonically covered by

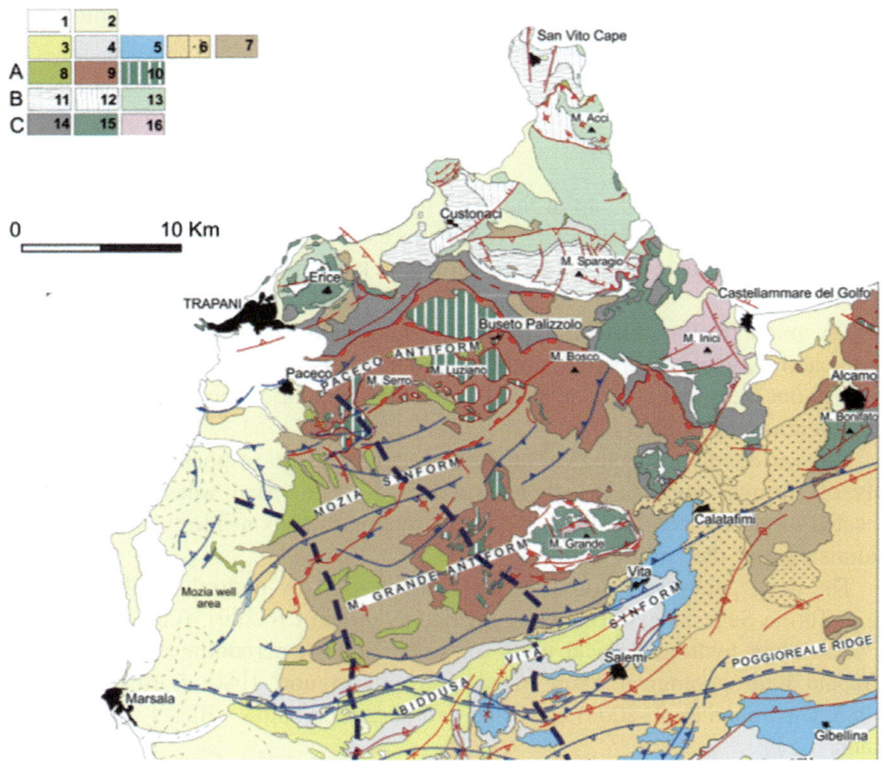

Fig. 12.2 Structural geological map of the Trapani area (modified from Catalano and Lo Cicero 1998). (1) Recent deposits; (2) Clastic carbonates, Pleistocene–Late Pliocene; (3) Marls and calcarenites, Pliocene; (4) Marly calcilutites (‘*Trubi*’), Early Pliocene; (5) evaporites, Messinian; (6) reef carbonates, Early Messinian; Terravecchia Fm., Early Messinian–Late Tortonian. (**A**) *Pre-Panormide units*: (7) marly arenaceous member (Langhian—Early Tortonian); (8) biocalcarenites and arenaceous strata (Mischio), Early Miocene; (9) calcarenites, shales and glauconitic quarzarenites, Oligocene; (10) marls and pelagic calcilutites with biocalcarenites, Early Cretaceous–Early Oligocene. (**B**) *Panormide units:* (7) marly arenaceous member, Langhian—Early Tortonian, glauconitic Calcarenites, Early Miocene; (11) shelf to slope calcarenites and calcilutites, Early Cretaceous—Oligocene; (12) calcilutites and calcarenites, Jurassic—Early Cretaceous; (13) platform carbonates and dolomites, Trias—Lias. (**C**) *Trapanese units*: (14) marly arenaceous member, Langhian—Early Tortonian; glauconitic calcarenites, Early Miocene; (15) pelagic platform deposits, Jurassic–Oligocene; (16) platform carbonates and dolomites, Trias—Lias

the Panormide units. These units, deformed at a large radius of curvature, are cut by reverse faults and/or overthrusts.

Tectonically superimposed on the Trapanese units we find the units that derive from the deformation of the Pre-Panormide domain, consisting of successions of both basin, margin and slope (sector between Paceco and Buseto Palizzolo). These are units that have been detached from their original Mesozoic substrate and are made up of successions with lithologies of prevalently ductile behaviour that have given rise to very marked folded structures and numerous small tectonic scales stacked to form imbricated structures.

12.2 Hydrogeology

The basin is mainly made up of the Trias limestone and dolomitic limestone overthrusting on the Tertiary deposits. These successions, consisting of carbonate and carbonate-siliciclastic complexes, fractured and karstified, are the site of important aquifers.

The main fracture networks, some of which identify areas of greater relative permeability, while others probably limit the transfers between adjacent water bodies without creating real underground watersheds, condition the underground water circulation in carbonate soils.

The various hydrogeological structures identified are responsible for the supply of springs, some of which have a low flow rate, and water wells, mostly located along the coastal strips, whose extraction is responsible for the increased salinisation of the coastal aquifers.

The deep geological structures are characterised by hydrothermal circulation (e.g. Terme Segestane, Terme Gorga, Erice, contrada Crociferi, Balata—Inici).

The main hydrogeological units (or water bodies) pertaining to the Monti di Trapani hydrogeological basin (Cusimano and Liguori 1980; Bartolomei et al. 1983; Cusimano et al. 2002) are listed and described below (Figs. 12.3, and 12.4; Table 12.1):

- **Mt Ramalloro–Mt Inici hydrogeological unit**: it consists essentially of carbonate and terrigenous carbonate deposits of the Upper Triassic—Miocene (Trapani Domain) and represents the main hydrostructure of the Trapani Mountains. This hydrostructure is tectonically covered by the hydrostructures belonging to the Panormide and Pre-panormide domains. The final water outlet of this unit is represented by the sea and the Pleistocene plain of Castellammare del Golfo. Transcurrent lines along the course of the Fiume Caldo appear to be responsible for hydrothermal vents (Terme Segestane and Terme Gorga), including those at Mount Inici, where several water wells with highly mineralised hot water fall (Cusimano et al. 1986). This unit is characterised by the presence of thermal groundwater, the most known are Segestane and Gorga spas, with an output of 200 l/s and temperatures of 40–45°C.
- **Mt Erice hydrogeological unit:** In the westernmost sector of the Trapani Mountains lies the hydrogeological unit of Mount Erice (717 m asl). The basic water table, with a prevalent underground water discharge to the north, feeds the reduced Quaternary esplanades, while other water flows directly into the sea. A hydrothermal water table was intercepted near its south-western slopes (Pozzo Milo, depth 310 m), which would further confirm the existence of a large carbonate reservoir in the subsoil, endowed with its own impermeable terrigenous cover and affected by surface and deep water circuits originating cold and

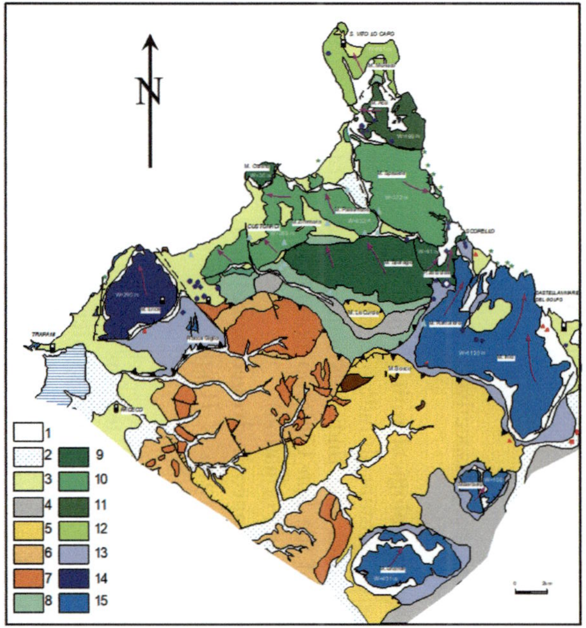

1) Debris; 2) Alluvial floods; 3) Bioclastic limestones and conglomerates, Plio-Pleistocene; 4) Sandy clays, sandstones, sands and conglomerates, Early Messinian-Upper Tortonian; (A) Pre-Panormide units: 5) Sandstones and pelites with intercalations of quartzarenitic and carbonate beds, Medium Oligocene-Upper Oligocene; 6) Clays, marls and marly limestones, Upper Eocene- Upper Miocene; 7) Marls and marly limestones with calpionella, calcilutites and marly limestones with planktonic foraminifera, biocalcites and biocalcirudites with macroforaminifera, Early cretaceous-Medium Miocene; (B) Panormide units: 8) Clays and sandy clays, Upper langhiano-Tortonian; 9) MT SPARAGIO-MT COFANO UNIT: Dolomites and dolomitic limestones, stromatolitic and loferitic limestones, nodular ammonitic limestones, ellipsactinic limestones and calpionellitic calcilutites, marly ammonitic and aptitic marls and calcilutites, rudist limestones with intercalations of volcanites, marly calcist limestones and calciscites with planktonic foraminifera, calcarenites and calcirutites with macroforaminifera, Upper Trias-Medium Miocene; 10) MT SPEZIALE-MT PALATIMONE: Dolomites and dolomitic limestones, stromatolitic and lateritic limestones, nodular ammonitic limestones, ellipsactinic limestones and calpionellitic calcilutites, marly ammonitic and aptitic marls and calcilutites, rudist limestones with intercalations of volcanites, marly calcist limestones and calciscites with planktonic foraminifera, calcarenites and calcirutites with macroforaminifera, Upper Trias-Medium Miocene; 11) MT ACCI UNIT: Dolomites and dolomitic limestones, stromatolitic and lateritic limestones with intercalations of volcanites, nodular ammonitic limestones, calpionellitic calcilutites, marls and marly calcilutites with radiolarians and aptic, marly calcilutites and calciscites with planktonic foraminifera, Upper Trias-Eocene; 12) MT MONACO UNIT: stromatolitic and lateritic dolomites, nodular ammonitic limestones, ellipsactinic limestones, rudist limestones, marly calcilutites and calciscites with planktonic foraminifera, calcarenites and calcirutites with macroforaminifera, Upper Trias-Medium Miocene; HYBLEAN-TRAPANESIAN UNIT: 13) Clays, marly clays and marls, Upper langhiano-Tortonian; 14) MT ERICE UNIT: Dolomites and dolomitic limestones, stromatolitic and lateritic limestones, calcilutites and calcisiltites with brachiopods and cephalopods, nodular ammonitic limestones, calpionellitic calcilutites, marny calcilutites and calcisiltites with planktonic foraminifera, Upper Trias-Upper Cretaceous; 15) MT INICI, MT BARBARO, ROCCA GIGLIO, MT GRANDE UNITS: Dolomites and dolomitic limestones, stromatolitic and lateritic limestones, nodular ammonitic limestones, calpionellitic calcilutites, marny calcilutites and calcisiltites with planktonic foraminifera, calcarenites and calcirutites with macroforaminifera, Upper Trias -Medium Miocene.

Fig. 12.3 Hydrostructural model of the Trapani Mountains (from Cusimano et al. 2002)

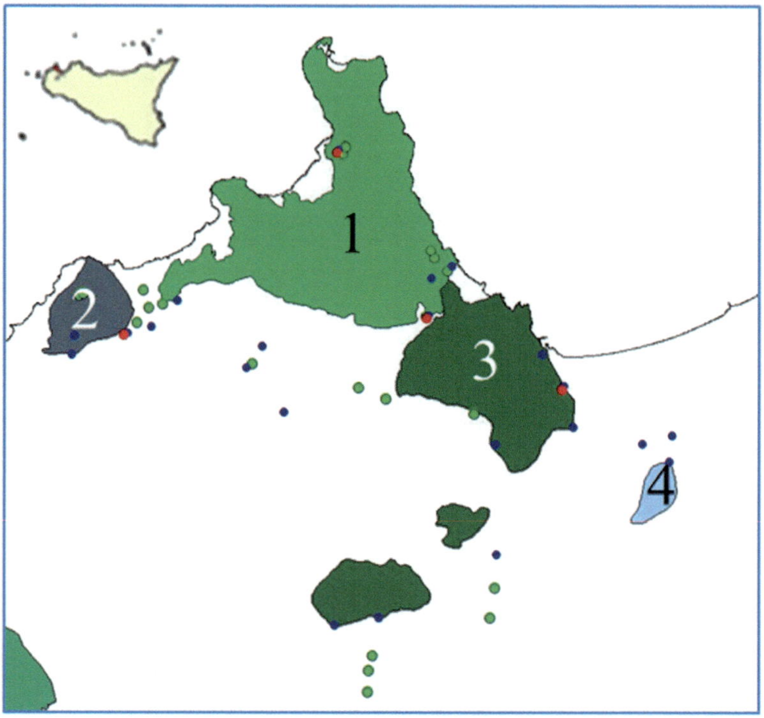

Fig. 12.4 Map of Trapani hydrological basin. Numbers identify water bodies: (1) Mt Sparagio–Mt Monaco; (2) Mt Erice; (3) Mt Ramarrolo–Mt Inici; (4) Mt Bonifato. The location of springs is also shown (green dots PRGA springs, blue dots INGV (2005), red dots indicate springs sampled for this work)

hydrothermal emergences respectively (Fancelli et al. 1991). The permeability is around 10^{-2}–10^{-4} m/s.

- **Mt Bonifato hydrogeological unit:** In the extreme eastern end of the Trapani Mountains (maximum altitude 825 m asl), to the E of the Fiume Freddo, directly S of Alcamo, one finds this unique outcrop of Trapani units buried under the Marls of the San Cipirello Formation and Tortonian fluvio-delithic deposits (Terravecchia Formation).
- **Mt Sparagio–Mt Monaco hydrogeological unit:** The hydrogeological unit, consists of carbonate deposits (Upper Triassic–Lower Miocene: Panormid domain).

The outcrops of the tardorogenic terrains contain several aquifers (the most productive in the arenaceous-conclomeratic intervals of the Terravecchia Formation), separated by impermeable or semi-permeable horizons. The Messinian evaporites, on the other hand, give rise to selenitic springs, used for irrigation purposes (Figs. 12.3, and 12.4).

Table 12.1 Description of Trapani Mts water bodies

Water body	Hydrogeological complex	Mouton hydrogeological complex	Description	PTA hydrogeological complex	Type of aquifer
Mt Erice	CA limestone	Limestone complex	Carbonate and terrigeous carbonate deposits (Trapanese Domain)	Limestone complex	Multifaceded aquifer fractured and karstified carbonate rocks
Mt Bonifato	CA limestone	Limestone complex	Carbonate and terrigeous carbonate deposits (Trapanese Domain)	Limestone complex	Multifaceded aquifer fractured and karstified carbonate rocks
Mt Sparagio–Mt Monaco	CA limestone	Limestone complex	Meso-Cenozoic carbonate succession with carbonate platform, margin and scarp terrigenous features (Panormide Domain)	Limestone complex	Multifaceded aquifer fractured and karstified carbonate rocks
Mt Erice	CA limestone	Limestone complex	Carbonate and terrigeous carbonate deposits (Trapanese Domain)	Limestone complex	Free aquifer fractured and/or karstified limestones

The Plio-Pleistocene sandy-calcarenitic and conglomeratic deposits are characterised by the presence of multi-water table aquifers, as are the alluvial deposits of the beds of current watercourses.

12.3 Hydrogeochemical Characterisation

In this basin, for water samples have been collected to represent two water bodies. The water temperature varies between 18 and 19 °C, the conductivity between 455 and 800 μS/cm (at 20 °C) and the pH ranges just around 7.5 (Table 12.2).

In the Langelier–Ludwig classification diagram, the waters are classified as bicarbonate-alkaline-earthy, and the waters of the Fico spring as chloride-sulfate-alkaline-earthy (Fig. 12.5).

Table 12.2 Chemical and isotopic composition of the sampled points

	T	EC	pH	Eh	Na	K	Ca	Mg	Cl
Fico	18.2	795	7.6	338	2.36	0.24	3.68	2.25	2.73
Inici	19.4	349	7.45	150	1.23	0.02	3.29	0.24	1.49
Menta	18.0	469	7.48	237	1.55	0.08	4.28	0.52	1.32
Macari	19.4	455	7.82	198	0.95	0.02	1.86	2.51	1.18

	NO_3	SO_4	HCO_3	δ^2H	$\delta^{18}O$	WB
Fico	0.669	1.12	3.91	−27	−6.27	MR-MI
Inici	0.15	0.38	3.10	−39	−6.83	MS-MM
Menta	0.734	0.47	3.85	−34	−6.09	MS-MM
Macari	0.008	0.28	3.90	−34	−6.45	MS-MM

Analytical data are expressed as meq/L, EC in μS/cm and δ^2H and $\delta^{18}O$ in ‰ versus SMOW. WB is for water body. MR-MI is for Mt Ramalloro–Mt Inici, MS–MM is Mt Sparagio–Mt Monaco water bodies

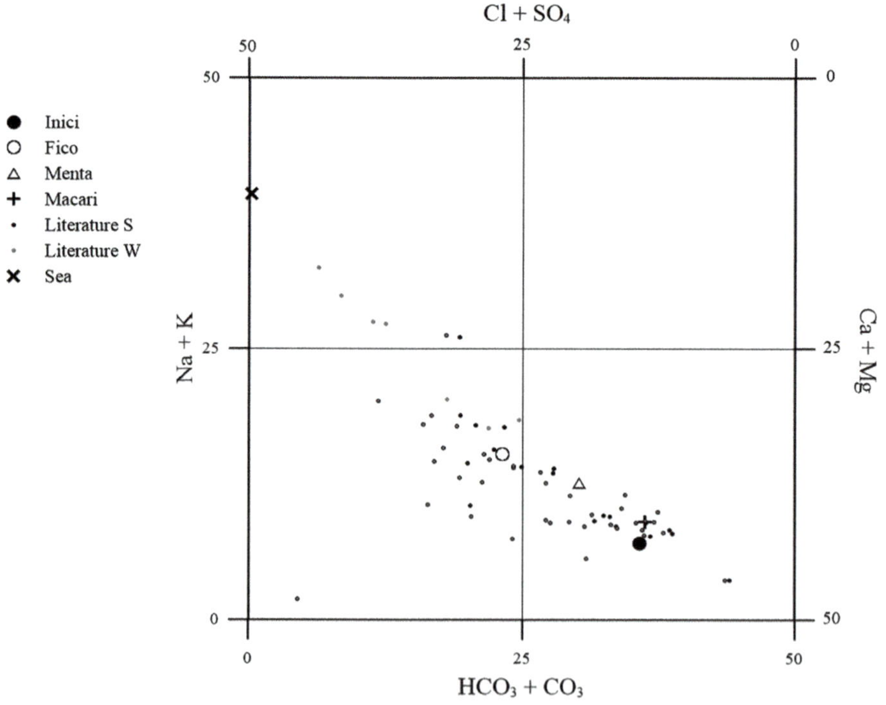

Fig. 12.5 Langelier–Ludwig classification diagram. The springs plot in the bicarbonate-alkaline-earthy and chlorine-sulphate-alkaline-earthy quadrants. Literature data (INGV 2005) are shown for comparison; S: spring, W: well

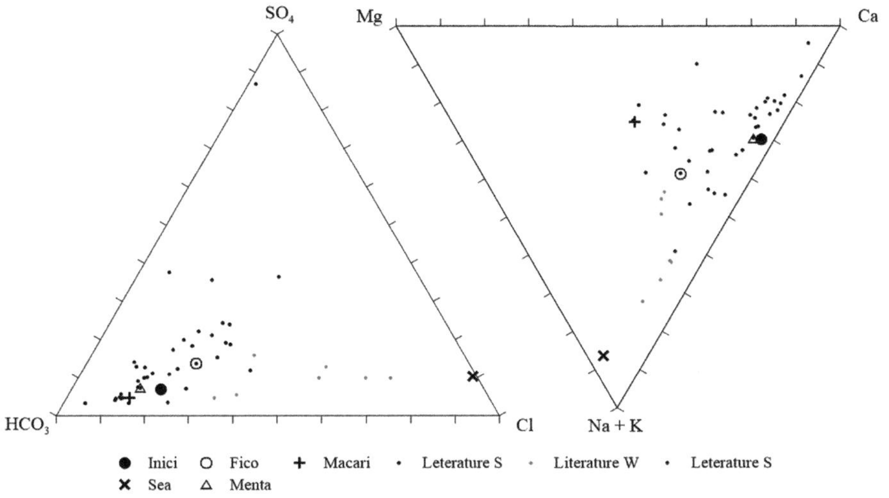

Fig. 12.6 Anions (left) and cations (right) ternary diagrams for Trapani Mountains catchment area. Literature data are shown for comparison (INGV 2005). S: Spring, W: Wells

The main processes governing the content of dissolved species in groundwater in the area under examination are the interaction of meteoric water with atmospheric particulate and with the rocks that make up the aquifer (Liotta 2004). In Western Sicily, the significant water bodies are set on the carbonate reliefs.

The geochemical characteristics of the groundwater reflect this peculiarity, presenting as evidenced by the triangular diagram in Fig. 12.6 a conspicuous calcium and magnesium content. The increase in sulphates, not strictly related to mixing with the marine end member, supports the hypothesis of leaching of an evaporitic gypsum origin. Furthermore, it has been observed that the amount of dissolved bicarbonates is closely related to the presence of carbon dioxide in the water table. The amount of dissolved CO_2 is not homogeneously distributed in the Trapani area and often reflects the spatial distribution of tectonic structures, as has been observed in the areas of Mt. Erice and Mt. Inici (Liotta 2004). In the northern area, the high permeability of the aquifers and the presence of karst circuits favor marine intrusion processes. This process can be identified in Figs. 12.5 and 12.6 by the arrangement of some of the samples towards the representative seawater end member.

12.4 Isotopic Characteristics of the Catchment Area

The isotopic composition of sampled waters plot between the GMWL and MMWL and above the LMWL. One point (Fico) falls on the GMWL line. The LMWL for the Trapani basins was determined by Liotta et al. (2013) as sector 1 (see Chap. 1,

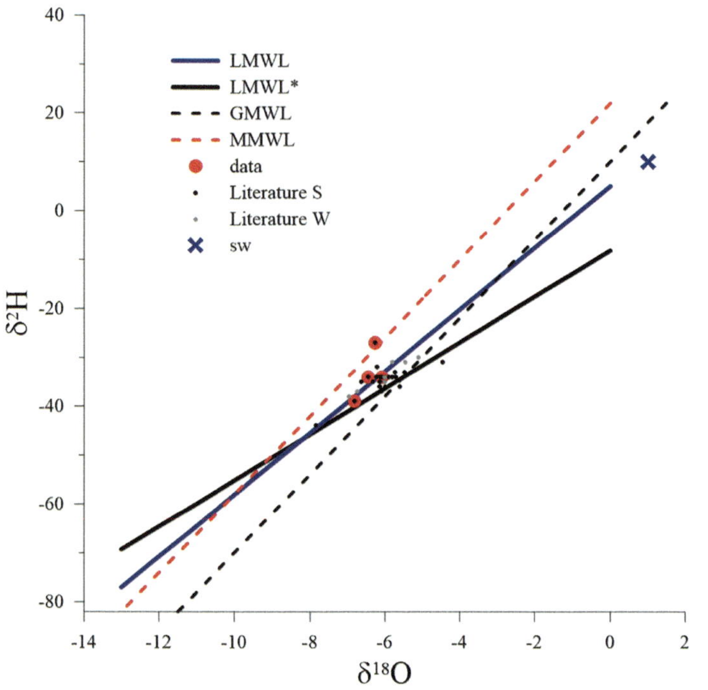

Fig. 12.7 δ^2H–δ^{18}O (in ‰ versus SMOW) diagram of spring waters. The LMWL is from Liotta et al. (2013), LMWL* is from Liotta et al. (2006). The GMWL and MMWL are shown for comparison. Literature data from INGV (2005). S: spring, W: well

Table 1.7). The LMWL is $\delta^2H = 6.31\ \delta^{18}O + 5$. The LMWL*, determined by Liotta et al. (2006) for the northwestern Sicily, is $\delta^2H = 4.7\ \delta^{18}O - 8.2$ (Fig. 12.7).

The weighted average composition of meteoric waters and the general trend of all water bodies show that the aquifers are generally fed by water from higher elevations, but undergo processes such as: local meteoric water input, seawater contamination and evaporation processes that alter their original composition.

The vertical oxygen isotopic gradient is −0.15‰/100 m, similar to that estimated by Liotta et al. (2006) for western Sicily (−0.18‰/100 m) (Table 1.8 relative to sector 1). In the plot in Fig. 12.8, springs seem to be fed by rain falling at about the same altitude as each spring. Unfortunately, the isotopic gradient, in this case, does not agree with the hydrogeological evidences described above. The cause of this discrepancy is probably due to the ample area where rain gauges are located, not representative of the “small”, relatively speaking, Trapani Mts hydrogeological basin area.

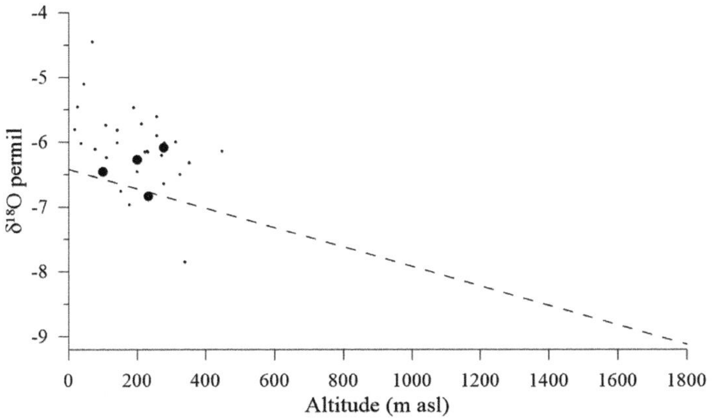

Fig. 12.8 Relationship between $\delta^{18}O$ and altitude

Bibliography and Recommended Further Reading

Abate B, Incandela A, Renda P (1990) Structural elements of the reliefs of Mount Erice and Rocca Giglio (Western Sicily). Rend Soc Geol It 13:99–102

Abate B, Di Maggio C, Incandela A, Renda P (1991) New data on the geology of the San Vito Lo Capo peninsula (north-western Sicily). Mem Soc Geol It Giornate in Memoria Di Leo Ogniben 47:15–25

Abate B, Incandela A, Renda P, Tramutol M (1996) Facies and sedimentary evolution of the 'Trapani Basin' (western Sicily) during the Oligocene-Miocene. Mem Soc Geol It 51:119–128

Abate B, Incandela A, Nigro F, Renda P (1998) Plio-Pleistocene strike-slip tectonics in the Trapani Mts. (NW Sicily). Boll Soc Geol It 117:555–567

Abate B, Incandela A (2004) Geological outline of the western end of Sicily. Atti del 4° Convegno di Speleologia della Sicilia, pp 25–35

Abate B, Di Maggio C, Incandela A, Renda P (1993) Geological Map of the Capo San Vito Mountains. Scale 1:25,000. Stabilimento Salomone, Roma

Agate M, D'Argenio A (1998) A review of NW Sicily offshore and Sicily Channel bedrock geology. Giorn Geol 60:205–217

Agate M, Catalano R, Infuso S, Lucido M, Mirabile L, Sulli A (1993) Structural evolution of the Northern Sicily Continental Margin during Plio-Pleistocene. In: Max MD, Colantoni P (eds) Geological development of the Sicilian-Tunisian platform. Proceedings of the international scientific meeting held at the University of Urbino, UNESCO Reports in Marine Science, vol 58, pp 25–30

Agate M, D'Argenio A, Di Maio D, Lo Iacono C, Lucido M, Mancuso M, Pepe F, Scannavino M (1998a) The Late Quaternary depositional sequence along the margin of north-western Sicily. Ext. Abs. In: Catalano R, Lo Cicero G (eds) Acts book, 79° Congr. Naz. S. G. I., Palermo (Italy) 21–23 September 1998, pp 32–34

Agate M, D'Argenio A, Di Maio D, Lo Iacono C, Lucido M, Mancuso M, Pepe F, Scannavino M (1998b) Sedimentary Dynamics of the Offshore of North-Western Sicily during the Late Quaternary. In: Catalano R, Lo Cicero G (eds) Guida alle escursioni vol. 1, 79° Congr. Naz. S. G. I., Palermo (Italy) 21–23 September 1998, pp 157–167

Agnesi V, Macaluso T, Pipitone G (1987) Role of deep gravitational deformations in the geomorphological evolution of the Scopello area (Trapani). Boll Soc Geol It 106:231–238

Agnesi V, Macaluso T, Monteleone S, Pipitone G, Sorriso-Valvo M (1989) Types and dynamics of deep gravitational deformations in relation to geological structures. The cases of Monte Genuardo and Scopello (western Sicily). Boll Soc Geol It 108:379–389

Agnesi V, Di Maggio C, Macaluso T, Madonia G, Rotigliano E (2004) Geomorphological outline of the San Vito Lo Capo peninsula (north-western Sicily). Atti del 4° Convegno di Speleologia della Sicilia, pp 37–47

Ambrosetti P, Bosi C, Carraro F, Ciaranfi N, Panizza M, Papani G, Vezzani L, Zanferrari A. (1983) Neotectonic Map of Italy (F. 5–6). C.N.R. P.F.G., neotettonica, L.A.C., Firenze

Bartolomei G, Celico I, Pecoraro A (1983) Hydrogeological scheme of north-western Sicily. Boll Soc Geol It 102:329–354

Bigi G, Cosentino D, Parotto M, Sartori R, Scandone P (1991) Structural model of Italy. (1/500.000). C.N.R. Progetto Finalizzato Geodinamica, Modello Strutturale tridimensionale, Firenze

Bommarito S, D'Angelo U, Vernuccio S (1992a) Geological Map of the Ummari Plate (F° 257 I NO, scale 1:25000). Nuova Graphicadue, Palermo

Bommarito S, D'Angelo U, Vernuccio S (1992b) Geological Map of the Segesta Plate (F° 257 I NE, scala 1:25000). Nuova Graphicadue, Palermo

Bommarito S, D'Angelo U, Vernuccio S (1992c) Geological Map of the Calatafimi plat (F° 257 I SE, scale 1:25000). Nuova Graphicadue, Palermo

Bommarito S, D'Angelo U, Vernuccio S (1992d) Geological Map of the Vita plat (F° 257 I SO, scale 1:25000). Nuova Graphicadue, Palermo

Broquet P, Mascle G (1972) The main stratigraphic and structural features of the northern part of the Trapani mountains (western Sicily). Ann Soc Géol Nord 92:139–146

Broquet P, Duée G, Mascle G (1975) A new element in favour of the allochthony of the Erice unit (western Sicily). C R Somm Soc Géol France 21(5):158–159

Broquet P, Duée G, Mascle G, Truillet R (1984) Recent alpine structural evolution in Sicily and its geodynamic significance. Rev Géol Dyn Et Géogr Phys 25(2):75–85

Carapezza M, Cusimano G, Liguori V, Alaimo R, Dongarrà G, Hauser S (1977) Introduction to the study of thermal springs on the island of Sicily. Boll Soc Geol It 96:813–836

Cassinis R (1967) Preliminaries on an investigation of freshwater efflux along the Sicilian coastline by means of salinity measurements at sea. Boll Geof Teor Appl 9:34

Catalano R, D'Argenio B (1978) An assay of palinspastic restoration across the Western Sicily. Geol Rom 17:145–159

Catalano R, Merlini S, Sulli A (2002) The structure of western Sicily, central Mediterranean. Petrol Geosc 8:7–18

Catalano R, D'Argenio B (1982) Geological outline of Sicily. In: Catalano R, D'Argenio B (eds) Guida alla geologia della Sicilia occidentale, Guide geologiche regionali, Mem. Soc. Geol. It., Suppl. a. XXIV, pp 9–41

Catalano R, Lo Cicero G (eds) (1998) Sicily, a natural laboratory in the Mediterranean. Structures, seas, resources and risks. 79° Congr. Soc. Geol. It., Palermo 18–20 settembre 1998, Guida alle escursioni, vol. 1, La Sicilia Occidentale, p 185

Catalano R, Franchino A, Merlini S, Sulli A, Agate M, Basilone L (1998) Materials for the Understanding of the Deep Settlement of Central Western Sicily. In: Catalano R, Lo Cicero G (eds) La Sicilia Occidentale. Field Trip Guide of the 79° Congr. Naz. S. G. I., Mondello (Pa), pp 175–185

Cusimano G, Di Cara A, Liguori V, Marescalchi P (1982) Hydrostructural model of the 'Fiume Caldo-Castellammare del Golfo' Thermal area (Trapani). Rend Soc Geol It 5:31–34

Cusimano G, Liguori V (1980) A hydrogeological synthesis of western Sicily (Italy). 26° Congrés Géologique Intern. Paris, 7–17 Juillet 1980, V. III, Résumés, Section 15 Hydrogéologie, p 1127

Cusimano G, Di Cara A, Marescalchi P (1986) Hydrogeological study of the Fiume Caldo-Castellammare del Golfo thermal area (Trapani). Atti del Congr. Intern.: "Igiene e medicina termale e ambientale", Castellammare di Stabia (Napoli), 24–26 Ott. 1986; II Sessione: Idrogeologia applicata al termalismo; suppl. al n. 2 della Rivista "Medicina Clinica e Termale", pp 180–189

Cusimano G, Frías Forcada A, Gatto L, Incandela A. (2002) Hydrostructural layout of the Trapani mountains (north-western Sicily) and evaluation of water resources. Atti del 4° Convegno di Speleologia della Sicilia, pp 117–124

D'Angelo U, Giorgianni A, Giunta G, Nigro F, Vernuccio S (1997) Neotectonic observations on the Capo San Vito Peninsula (NW Sicily). Il Quaternario 10(2):349–354

D'Angelo U, Ruggieri G, Unti M, Vernuccio S (1980) Neotectonics of Sheets 248 (Trapani), 249 (Palermo), 250 (Bagheria), 258 (Alcamo), 259 (Termini Imerese) e 266 (Sciacca). In: Contributions to the realisation of the Neotectonic Map of Italy, pubbl. n. 356, CNR-P. F. Geodinamica, 203–236, Roma

Di Maggio C, Incandela A, Masini F, Petruso D, Renda P, Simonelli C (1999) Shoreline Fluctuations, Biochronology of Continental Deposits and Neotectonics in North-Western Siclia (San Vito Lo Capo-Trapani Peninsula). A First Interdisciplinary Approach. Il Quaternario 12(1):25–49

Fancelli R, Monteleone S, Nuti S, Pipitone G, Rini S, Taffi L (1991) New hydrogeological and geothermal knowledge in western Sicily. Geol Appl Idrogeol 26:165–206

Favara R, Grassa F, Inguaggiato S, D'Amore F (1998) Geochemical and hydrogeological characterization of thermal springs in Western Sicily, Italy. J Volcanol Geotherm Res 84:125–141

Favara R, Grassa F, Inguaggiato S, Valenza M (2001) Hydrogeochemistry and stable isotopes of thermal springs: earthquake-related chemical changes along Belice Fault (Western Sicily). Appl Geochem 16:1–17

Gaffurini U (1954) Tertiary soil residues on the summit of Mount Bonifato di Alcamo. Boll Serv Geol It 76:277–282

Ghisetti F, Vezzani L (1984) Thin-skinned deformations of the Western Sicily Thrust Belt and relationships with crustal shortening: mesostructural data on the Mt. Kumeta-Alcantara fault zone and related structures. Boll Soc Geol It 103:129–157

Giannotti A (1958) Two Upper Jurassic facies in Sicily. Rev Micropaleontol 1(1):38–51

Giunta G, Liguori V (1970) Geology of the Capo San Vito Peninsula (North-Western Sicily). Lav Ist Geol Univ Palermo 9:21

Giunta G, Liguori V (1972) Geology of the North-Western extremity of Sicily. Riv Min Sic 136–138:165–226

Giunta G, Liguori V (1973) Paleotectonic Evolution of North-Western Sicily. Boll Soc Geol It 92:903–924

Incandela A (1995) Structural stratigraphic lineaments of the north-western tip of Sicily and the islands of Favignana and Levanzo (Egadi Archipelago). 216 pp. Phd Thesis

Liotta M, Grassa F, D'Alessandro W, Favara R, Gagliano Candela E, Pisciotta A, Scaletta C (2013) Isotopic composition of precipitation and groundwater in Sicily, Italy. Appl Geoch 34:199–206

Liotta M (2004) Geochemical processes governing groundwater composition in North-Western Sicily: isotopic model and water-rock interaction. Phd thesis—University of Palermo

Malatesta A (1955) Preliminary results of the survey in Sicily and the Egadi Islands. Boll Serv Geol It 215:517–521

Martire L, Pavia G, Pochettino M, Cecca F (2000) The middle-upper jurassic of montagna grande (Trapani): age, facies and depositional geometries. Mem Soc Geol It 55:219–225

Mascle G (1979) Geological study of the Sicani Mountains. Riv It Paleont Strat Mem 16:1–431

Mascle G (1970) Geological sketch of Western Sicily. In: Alvarez W, Gohrbandt KHA (eds) Geology and history of Sicily, 231-243, Petr. Expl. Soc. of Libya, Tripoli

Mauz, B, Renda, P (1995) Tectonic features at the NW-coast of Sicily (Gulf of Castellammare). Implications for the Plio-Pleistocene structural evolution of the southern Tyrrhenian continental margin. In: Cello G, Deiana G, Pierantoni PP (eds) Geodinamica e tettonica attiva del sistema Tirrenno-Appennino. Studi Geol. Camerti, Spec Issue 1995/2, pp 343–349

Mauz B, Renda P (1996) Geological Map of the Plain of Partitico and Castellamare del Golfo (scale 1:25000). Stabilimento Salomone, Roma

Mauz B, Renda P (1991) Tectonosedimentary Evolution of the Plio-Pleistocene Basin of Castellammare del Golfo (Northwestern Sicily). Mem Soc Geol It 47:167–180

Montanari M (1969) On the tertiary section of Monte Bonifato (western Sicily). Riv Ital Paleont 75(2):365–416

Montanari L (1987) Stratigraphic-paleogeographical features of Sicily during the Alpine cycle. Mem Soc Geol It 38:361–406

Montone P, Amato A, Frepoli A, Mariucci MT, Cesaro M (1997) Crustal stress regime in Italy. Ann Geof 40(3):741–757

Nairn AEM, Nardi G, Gregor CB, Incoronato A (1985) Coherence of the Trapanese units during tectonic emplacement in Western Sicily. Boll Soc Geol It 104:267–272

Nigro F, Renda P (1999) Plio-Pleistocene Wrench Tectonics in the Western Sicily Chain. ASGP 69:99–112

Nigro F, Renda P (2000) A model of tectono-sedimentary evolution of the Sicilian Neogene foreland. Boll Soc Geol It 119:667–686

Nigro F, Renda P (2001) Oblique-slip thrusting in the Maghrebide chain of Sicily. Boll Soc Geol It 120:187–200

Nigro F, Renda P (2002) Forced mode dictated by foreland fault-indenter shape during oblique convergence: the Western Sicily mainland. Boll Soc Geol It 121:151–162

Nigro F, Renda P (2004) The contribution of the pre-existing structures in the mountain belt evolution: the example of the negative inversion in Northern Sicily. Boll Soc Geol It 123:175–187

Nigro F, Renda P, Arisco G (2000) Recent Tectonics in North-Western Sicily and the Egadi Islands. Boll Soc Geol It 119:307–319

Ruggieri G (1966) First Research Results on the Tectonics of Western Sicily. Geol Romana 5:453–456

Schutte KG (1978) Crustal structure of southern Italy. In: Closs H, Roeder D, Schmidt K (eds) Alps, apennines, hellenides, Stuttgart, pp 315–321

Trimaille H (1982) Geological study of the Trapani Basin (Sicily, Italy). PhD. Thesis, Univ. Franche-Comté, p 177

Wendt J (1971) Geology of Mount Erice (Trapani Province, western Sicily). Geol Rom 10:53–76

GPSR Compliance

The European Union's (EU) General Product Safety Regulation (GPSR) is a set of rules that requires consumer products to be safe and our obligations to ensure this.

If you have any concerns about our products, you can contact us on ProductSafety@springernature.com

In case Publisher is established outside the EU, the EU authorized representative is:

Springer Nature Customer Service Center GmbH
Europaplatz 3
69115 Heidelberg, Germany

Batch number: 10726135

Printed by Printforce, the Netherlands